Metaheuristic Approaches for Optimum Design of Reinforced Concrete Structures:

Emerging Research and Opportunities

Aylin Ece Kayabekir
Istanbul University-Cerrahpaşa, Turkey

Gebrail Bekdaş
Istanbul University-Cerrahpaşa, Turkey

Sinan Melih Nigdeli
Istanbul University-Cerrahpaşa, Turkey

A volume in the Advances in
Chemical and Materials Engineering
(ACME) Book Series

IGI Global
DISSEMINATOR OF KNOWLEDGE

Published in the United States of America by
IGI Global
Engineering Science Reference (an imprint of IGI Global)
701 E. Chocolate Avenue
Hershey PA, USA 17033
Tel: 717-533-8845
Fax: 717-533-8661
E-mail: cust@igi-global.com
Web site: http://www.igi-global.com

Library of Congress Cataloging-in-Publication Data

Names: Kayabekir, Aylin Ece, 1993- author. | Bekdas, Gebrail, 1980- author. | Nigdeli, Sinan Melih, 1982- author.
Title: Metaheuristic approaches for optimum design of reinforced concrete structures : emerging research and opportunities / Aylin Ece Kayabekir, Gebrail Bekdaş, and Sinan Melih Nigdeli.
Description: Hershey PA : Engineering Science Reference, [2020] | Includes bibliographical references and index. | Summary: "This book explores the use of metaheuristic approaches in the optimum design of reinforced concrete structures"-- Provided by publisher.
Identifiers: LCCN 2019042069 (print) | LCCN 2019042070 (ebook) | ISBN 9781799826644 (hardcover) | ISBN 9781799826651 (paperback) | ISBN 9781799826668 (ebook)
Subjects: LCSH: Buildings, Reinforced concrete--Design and construction--Data processing. | Reinforced concrete construction--Mathematics. | Metaheuristics.
Classification: LCC TA683.2 .K39 2020 (print) | LCC TA683.2 (ebook) | DDC 624.1/8341--dc23
LC record available at https://lccn.loc.gov/2019042069
LC ebook record available at https://lccn.loc.gov/2019042070

This book is published in the IGI Global book series Advances in Chemical and Materials Engineering (ACME) (ISSN: 2327-5448; eISSN: 2327-5456)

British Cataloguing in Publication Data
A Cataloguing in Publication record for this book is available from the British Library.

Advances in Chemical and Materials Engineering (ACME) Book Series

ISSN:2327-5448
EISSN:2327-5456

Editor-in-Chief: J. Paulo Davim University of Aveiro, Portugal

MISSION

The cross disciplinary approach of chemical and materials engineering is rapidly growing as it applies to the study of educational, scientific and industrial research activities by solving complex chemical problems using computational techniques and statistical methods.

The **Advances in Chemical and Materials Engineering (ACME) Book Series** provides research on the recent advances throughout computational and statistical methods of analysis and modeling. This series brings together collaboration between chemists, engineers, statisticians, and computer scientists and offers a wealth of knowledge and useful tools to academics, practitioners, and professionals through high quality publications.

COVERAGE

- Ceramics and glasses
- Ductility and Crack-Resistance
- Thermo-Chemical Treatments
- Multifuncional and Smart Materials
- Statistical methods
- Industrial Chemistry
- Computational methods
- Sustainable Materials
- Fracture Mechanics
- Polymers Engineering

IGI Global is currently accepting manuscripts for publication within this series. To submit a proposal for a volume in this series, please contact our Acquisition Editors at Acquisitions@igi-global.com or visit: http://www.igi-global.com/publish/.

The Advances in Chemical and Materials Engineering (ACME) Book Series (ISSN 2327-5448) is published by IGI Global, 701 E. Chocolate Avenue, Hershey, PA 17033-1240, USA, www.igi-global.com. This series is composed of titles available for purchase individually; each title is edited to be contextually exclusive from any other title within the series. For pricing and ordering information please visit http://www.igi-global.com/book-series/advances-chemical-materials-engineering/73687. Postmaster: Send all address changes to above address.

Titles in this Series

For a list of additional titles in this series, please visit:
https://www.igi-global.com/book-series/advances-chemical-materials-engineering/73687

Properties, Techniques, and Applications of Polyaniline (PANI) Thin Films Emerging Research and Opportunities
Subhash Chander (Indian Institute of Science Education and Research, Mohali, India) and Nirmala Kumari Jangid (Banasthali University, Tonk, ndia)
Engineering Science Reference • © 2020 • 236pp • H/C (ISBN: 9781522598961) • US $165.00

Applications and Techniques for Experimental Stress Analysis
Karthik Selva Kumar Karuppasamy (Indian Institute of Technology, Guwahati, India) and Balaji P.S. (National Institute of Technology, Rourkela, India)
Engineering Science Reference • © 2020 • 269pp • H/C (ISBN: 9781799816904) • US $245.00

Design of Experiments for Chemical, Pharmaceutical, Food, and Industrial Applications
Eugenia Gabriela Carrillo-Cedillo (Universidad Autónoma de Baja California, Mexico) José Antonio Rodríguez-Avila (Universidad Autónoma del Estado de Hidalgo, Mexico) Karina Cecilia Arredondo-Soto (Universidad Autónoma de Baja California, Mexico) and José Manuel Cornejo-Bravo (Universidad Autónoma de Baja California, Mexico)
Engineering Science Reference • © 2020 • 429pp • H/C (ISBN: 9781799815181) • US $255.00

Handbook of Research on Developments and Trends in Industrial and Materials Engineering
Prasanta Sahoo (Jadavpur University, India)
Engineering Science Reference • © 2020 • 524pp • H/C (ISBN: 9781799818311) • US $295.00

Nanocomposites for the Desulfurization of Fuels
Tawfik Abdo Saleh (King Fahd University of Petroleum and Minerals, Saudi Arabia)
Engineering Science Reference • © 2020 • 358pp • H/C (ISBN: 9781799821465) • US $225.00

IGI Global
DISSEMINATOR OF KNOWLEDGE

701 East Chocolate Avenue, Hershey, PA 17033, USA
Tel: 717-533-8845 x100 • Fax: 717-533-8661
E-Mail: cust@igi-global.com • www.igi-global.com

Table of Contents

Preface

The goal of an engineer is to complete a design in a best way considering safety, esthetics and practicability with the minimum cost. By the shortening of earth resources, the usage of construction materials must be at the minimum required level. For that reason, optimum design of structural members is important in the present time and it will be more important in the future.

Generally, several decisions in the design of civil structures are done according to the experience of engineers. These decisions include material types, member dimensions and amount of additional materials. Several trials for optimum design problems have high number of constraints related with design variables. The best way is to use numerical iterations. Metaheuristic algorithms are the best option to consider design constraints defined in the design regulation codes of reinforced concrete structures which is a composite of two materials with different behavior; concrete and steel.

Reinforced concrete (RC) structures are one of the major type of structures. Since the tensile capacity of concrete is low, the structural members are reinforced with steel bars. According to the design regulation codes, the rules of safe design are given. The optimization problem is to find the best design (section dimension, material type and amount of reinforcement) with the minimum cost providing the design constraints (design formulation considering loading of structure).

Metaheuristic methods inspired by natural phenomena are the one of the best options to consider design constraints by combining the analyses of formulation of reinforced concrete structures with an iterative numerical algorithm using several convergence options of random generation of candidate design solutions.

The current book contains several metaheuristic algorithms and the design of several types of structural members. Also, retrofit applications and seismic design issues are considered for readers from earthquake zone areas.

Additionally, the current place of science is given in a state of art review chapter. Computer codes are also provided in the book.

The economy becomes the most important issue in current days. A design providing safety and esthetics is not enough for an engineering design. The nonlinear behavior of optimum design of structures preventing the usage of mathematical methods. In that case, the use of metaheuristic algorithms plays an important role.

To make an optimum design of RC member via metaheuristic method, a computer code must be generated. In this code, the calculation of internal forces (analysis of structure), the calculation of different types of stresses in several sections and the formulations of algorithms must be provided. In this book project, several computer codes are provided and step by step methodologies are given.

Furthermore, the book is a great source for academicians since the newly generated optimum design studies are generally employed metaheuristic approaches. Also, optimum design of structures is given as a lecture in universities in graduate and undergraduate levels in civil engineering. The book is also a great source for design engineers.

The applications of metaheuristic algorithms become widespread with advancing technology in the optimum design of engineering problems. Also, all parts of the world must protect resources and the structures must be cheaper while providing all incomes to the residents. Especially, reinforced concrete structures have highly constrained optimum design and the use of metaheuristics is important in this subject. In many countries, in the field of educations such as universities, research centers etc. are conveyed in various lectures or courses for undergraduate and graduate students.

In addition, the book is internationally acceptable. The constraints are handled according ACI-318 - Building Code Requirements for Structural Concrete of American Concrete Institute, which is internationally accepted or easy to adapt to local design codes.

The first chapter of the book includes a review survey about optimization of reinforced concrete (RC) is presented together with an introduction part about metaheuristic algorithms. This introductory part includes the necessity of using metaheuristics in engineering problems with nonlinear behavior (resulting from design constraints) and development of optimization of structure starting from early attempts of Galilei Galileo.

The second chapter is a general one, in which the brief history and exploration of metaheuristic algorithms can be found. Before the brief history and definitions of various algorithms, the reasons of using metaheuristics for optimization of RC structures are listed as seven reasons. Then, the brief history is explained in five periods; namely the pre-theoretical period, the early period, the method-centric period, the framework-centric period and scientific period. The essential source of metaheuristic is the human mind and the use of heuristic starts by the born of a baby. The first development of optimization algorithms started after the World War II (1940s), but the human mind was the source of heuristics before the 1940s. In the early period, the principles of heuristics such as analogy, induction and auxiliary problem are mentioned. The true start of evolutionary algorithms was developed in the method-centric period and the major features of new generation algorithms were also mentioned for the classical algorithms. In the framework-centric period, the framework or metaphor used in algorithms became more important than the method. In the future (the scientific period), metaheuristics will remain important and the solving of problems will be more essential than metaphor.

As explained in Chapter 2, the indexes used in optimization problems are design constants, design variables, objective functions and design constraints. A fundamental methodology of metaheuristic methods is presented. The methodology contains pre-optimization stage, analysis stage and optimization stage. The optimization stages of various algorithms are explained with the related flowcharts.

The other chapters include several optimization applications about RC members.

In Chapter 3, two optimum design methodology for design of RC beams using two different international design codes are presented. The computer code of one of the applications is given at the end of the chapter.

The fourth chapter is related to RC columns. For a basic optimization example, the written computer code according to the presented methodology is given. In this chapter, important considerations including slenderness effects on design of RC columns are also mentioned.

Foundation of RC structures is another important issue for optimization. After the two core elements of RC structures are mentioned and explained in the previous chapters, optimum design of RC footings is explained in Chapter 5. A computer code is also given in this chapter.

Frame structures which are the combination of beam and column elements must be considered together with the structural analysis part in optimization. This process is explained in Chapter 6.

Optimum design of RC retaining walls considering both geotechnical and structural design constraints is presented in Chapter 7, and an application code is presented as Appendix.

In Chapter 8, an optimum application for beams using carbon fiber reinforced polymers is presented by including a computer code. By including this chapter, a retrofit application for existing structures is also included in the book.

In Chapter 9, optimum post-tensioning application for axially symmetric cylindrical RC walls used as liquid tanks is presented. The last chapter includes a non-classical application and a special case is also included.

Chapter 1
Introduction and State-of-the-Art Review of Optimum Design of Reinforced Concrete Structures:
Overview for Optimum Design RC Structures

ABSTRACT

This first chapter of the book presents an introduction and review study. The necessity of optimization in engineering design is discussed. The nonlinear behavior of problems plays an important role in the usage of metaheuristic methods because of complexity resulting from design constraints considering safety and utilization rules. Design factors in analysis and design of structures are given. A brief history about optimization of structures is presented, including the first early attempts of Galilei Galileo. As the main scope of the book, the review of studies considering optimization of reinforced concrete (RC) structures and members via metaheuristic methods are given. The optimized RC members include beams, columns, slabs, frames, bridges, footings, shear walls, retaining walls, and cylindrical walls.

DOI: 10.4018/978-1-7998-2664-4.ch001

THE NECESSITY OF OPTIMIZATION IN ENGINEERING

Several issues must be handled when an engineering design is done. These important issues are about economy, safety, utilization and architecture. To verify a design as an engineered object, the consideration of one or several of these important issues is not enough. An engineering design must consider all these important issues to maintain the needs of individuals. In that case, an engineer has a heavy work to think all issues to find an excellent balance between them. Only experienced ones can handle all issues perfectly. To consider these issues, several measures must be generated, and these issues must be formulated in the design. Since these issues are related to each other, the design problems are generally highly nonlinear, and several methods must be developed, and these methods must be specific to the design problem. Because of non-linearity, these issues can be approximately handled by using conventional methods. Because of increasing demands and advantages of technology, today's needs are beyond an approximate solution. Several non-linearity may be hardly or approximately handled, or a classical solution cannot be found by without assumption of several factors. In that case, the experience may play a great role. For a robust design, the consideration of design stages can be iteratively done by employing an algorithm.

Civil engineering is one of the fundamental engineering disciplines and the products of civil engineers are directly effective on individuals since all reported issues such as economy, safety, utilization and architecture are extremely important in civil engineering designs. Civil engineering has a lot of applications due to all designs providing a living environment fall into the scope of civil engineering. In sub-disciplines of civil engineering such as structural engineering, structural mechanics, structural materials, construction management, geotechnical engineering, transportation engineering and hydraulics, optimization concerns many applications including water resources, superstructure and infrastructure projects, mass transportation and resources, traffic in transportation, stabilization of soil backfills and improvement of the soil.

Civil engineering designs suffers from the environmental conditions and the world has different geographic areas with different resources and differential risks of natural disasters. The most important issue in the design of civil engineering applications is safety, but the other issues are the index of the preferability.

For the safety of designs, the theory of structural mechanics and experimental investigations are formulized in several regulations. Although the theory and experiments are generally same at regulations, standardizations are different in several regions or different annexes are existing in different countries because of the difference of production quality of materials, workmanship quality and specific natural disasters in the region. The differences of regulations are combined with demands of people and social authorities including resources possibilities in a region. In that case, an optimum design for a region or an individual may not be the best acceptable design for another place. For that reason, these factors are considered as specific design constraints and design variable ranges.

The optimization of a product of mechanical engineering and a structure of civil engineering may show great differences in assigning optimum design results. To find the best optimum solutions, the use of continuous design variables is the best option, and a machine part or an object can be produced in a factory with precise dimension values. In a construction yard, the construction of the building with precise dimension cannot be provided. In that case, discrete variables can be used for structural design or the continuous values must be rounded to practical dimension. In that case, precise optimum design can never be used for structural design. Especially, reinforced concrete (RC) structures involves casting of concrete at site and placement of reinforcement with sizes found in local markets needs a specific assignment of design variables. For that reason, the size of reinforcing steel bars (rebar) must be selected from defined sizes by designer during optimization processes. Also, for steel structures, the profile sizes are produced according to fixed sizes of local markets. The optimum design of structures involves the best possible dimensions and design detail which is practical in construction.

The design rules in regulation are defined as design constraints and these constraints can be only found if the final dimension of design is known. Because of these constraints, the minimization of an objective function is not possible by using linear mathematical methods. For nonlinear methods, the existing of several types of design constraints (flexural capacity, shear capacity, minimum and maximum limits and ductility condition) and variables (cross-sectional dimension and amount of rebar with orientation for different types of stresses and sections) makes the optimization complex. In that case, numerical optimization methods using iterative stages are suitable for optimization.

For recent years, metaheuristic based optimum design methodologies have been developed for optimum design of civil engineering structures. Metaheuristic methods are the algorithms that formulize a process, happening or behavior from social life or nature, because every process, happening or behavior has a final aim like optimizing objectives in engineering problems. The details of several metaheuristic algorithms which are potentially used in engineering problems will be given in the Chapter 2.

ANALYSIS AND DESIGN OF STRUCTURES

The analysis of structures is done to find stresses in critical section of structural members. A structure is constructed by several members and each member has different stresses in mean of intensity and type. Also, a single member has varying stresses for different cross section. For economy, the design of different sections is separately done by considering several extremum sections.

The stresses of section occur because of internal forces and these forces are the ones holding various members of a structure. These internal forces may be axial force (N), shear force (V), flexural moment (M) and torsional moment (T). The most basic members used in structures are two-force members as shown in Figure 1. These members have only axial force and these members are straight.

The combination of two-force members with hinges generates truss structures. For hinges, rotation of joints is permitted. Also, all forces are concentrated at the joints. By using these assumptions, bending effects are neglected. In that case, members of truss structure are only under the effect of normal stress (σ), which is proportional to the area of the member (A) as shown in Figure 2.

Normal stresses on truss members can be compression or tension. For slenderness, long members are generally assigned as tension members. In that case, long members cannot be permitted for truss structures. For that

Figure 1. Two-force members

Figure 2. Member with an axial load

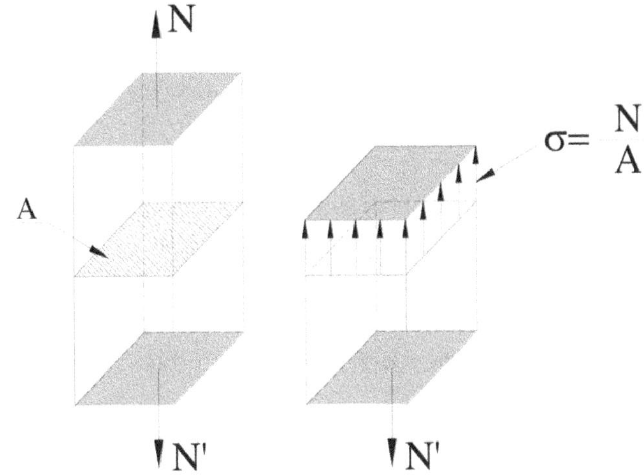

reason, the sections are also small for neglecting bending effects. In that case, a single material is used for truss members which are generally steel.

Since truss structures are used in roof tops, bridges towers ad space structures, the minimization of the total weight of the structure is essential and it is also directly related to the cost minimization. The optimization of truss structure is the most used application for metaheuristic algorithms. The main idea of optimization studies is to minimize the total weight of truss structure by optimizing areas of grouped structural members under design constraints such as maximum allowed stress and displacements. In several studies, the topology of truss structures has been also considered additional to sizing optimization. The employed metaheuristic algorithms for optimization of truss structures are Genetic Algorithm (GA) (Rajeev & Krishnamoorthy, 1992; Koumousis & Georgiou, 1994; Coello & Christiansen, 2000; Kelesoglu, 2007; Šešok & Belevičius, 2008; Toğan & Daloğlu, 2008; Li, 2015), Particle Swarm Optimization (PSO) (Schutte & Groenwold, 2003; Perez & Behdinan, 2007), Ant Colony Optimization (ACO) (Camp, 2004; Li, Huang, Liu, & Wu, 2007; Kaveh & Talatahari, 2009a), Teaching Learning Based Optimization (TLBO) (Degertekin & Hayalioglu, 2013; Camp & Farshchin, 2014; Dede & Ayvaz; 2015), Artificial Bee Colony (ABC) (Sonmez, 2011), Firefly Algorithm (FA) (Miguel, Lopez, & Miguel, 2013), Cuckoo Search (CS) (Gandomi, Talatahari, Yang, & Deb, 2013), Bat Algorithm (BA) (Talatahari & Kaveh, 2015), Big Bang Big Crunch (BB-BC) (Camp, 2007; Kaveh & Talatahari, 2009b; Hasançebi & Kazemzadeh Azad, 2014), Mine Blast Algorithm (MBA)

(Sadollah, Bahreininejad, Eskandar, & Hamdi, 2012) and Flower Pollination Algorithm (FPA) (Bekdaş, Nigdeli, & Yang, 2015).

The optimization of truss structures contains less design constraints than frame structures. Frame structures involve continuous connection of structural members without hinges and external loads are distributed to structural members as seen in Figure 3.

Because of strong winds and earthquake, frame structures are also subjected to lateral forces as shown as Figure 4. These time-varying forces and vibrations can change direction and intensity.

A plane frame member is shown as Figure 5 with the well-known positive directions. Even for a plane member, it is subjected to three types of internal forces such as N, V and M.

Because of the flexural moments, stress on a section vary according to distance to the neutral axis. In that case, the stresses are defined with the second moment of area (moment of inertia: I) as formulated as Eq. (1) and Eq. (2) for stress on a section shown as Figure 6.

$$\sigma_b = -\frac{My}{I} \tag{1}$$

$$\sigma_m = \frac{Mc}{I} \tag{2}$$

Figure 3. Frame structure with distributed loads (q)

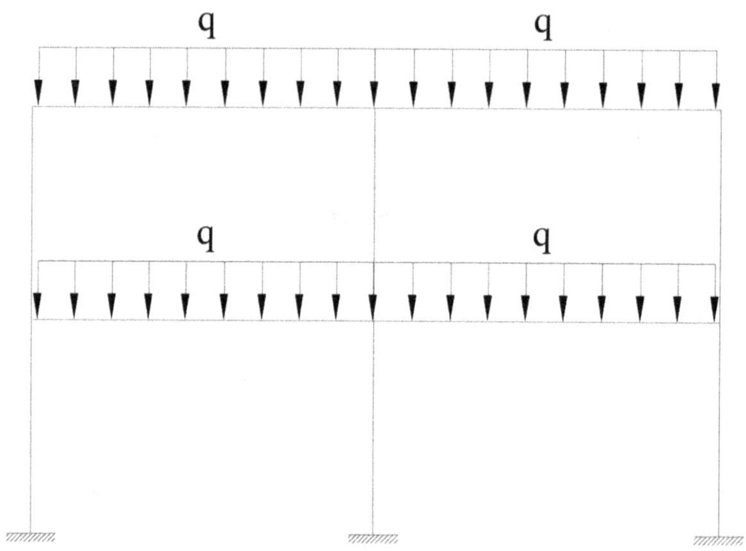

Figure 4. Frame structure with distributed loads (q)

Figure 5. A plane frame member

Figure 6. Member under flexural moment

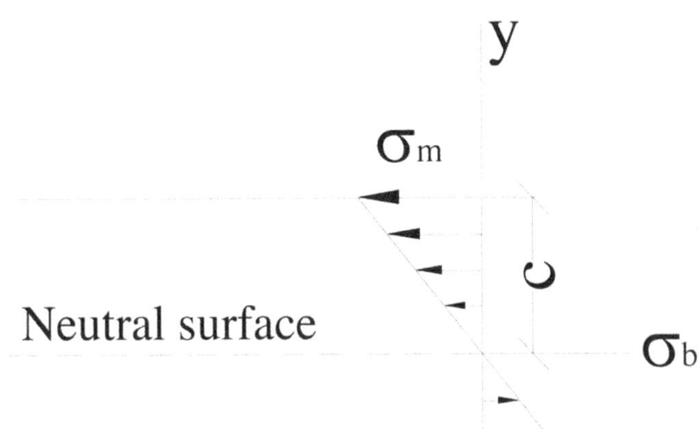

Because of the flexural moments, stress type (tension or compressive) is different in the edges of the cross section. By additional stress of an axial force, it can also change direction of the stress as shown in Figure 7.

Figure 7. Combined effect of axial load (N) and flexural moment (M)

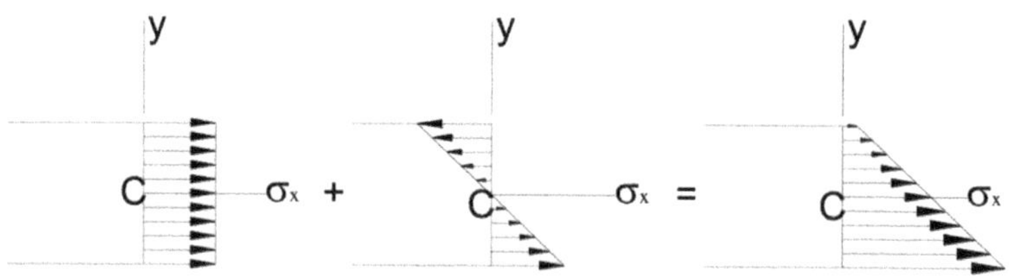

Additionally, shear stresses occur because of shear force. In figure 8, a transverse load shearing a section is shown and the shear stress (τ) distribution is shown as Figure 9.

Since a section can be under the effect of tensile and compressive stresses, material used in design must be a ductile material showing the same behavior and strength for tensile and compressive stresses. Using steel members is an option to use in structures. Several studies have been conducted for optimum design of steel frame structures (Degertekin, 2008; Saka, 2009; Toğan, 2012; Kaveh & Bakhshpoori, 2013; Camp, Bichon, & Stovall, 2005; Hasançebi, Çarbaş, Doğan, Erdal, & Saka, 2010).

Figure 8. A member with transverse loads

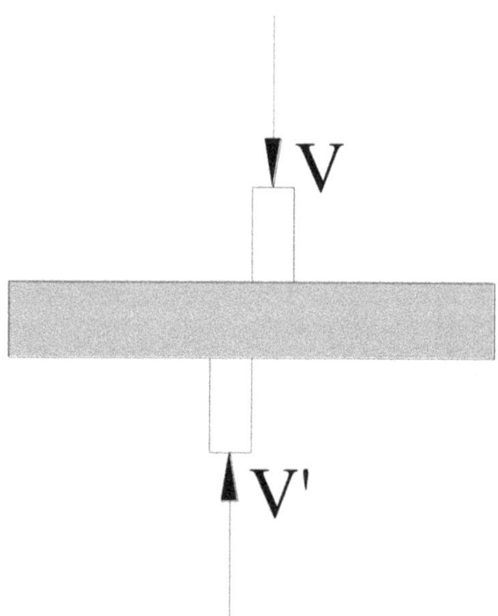

Figure 9. Shear stress distribution

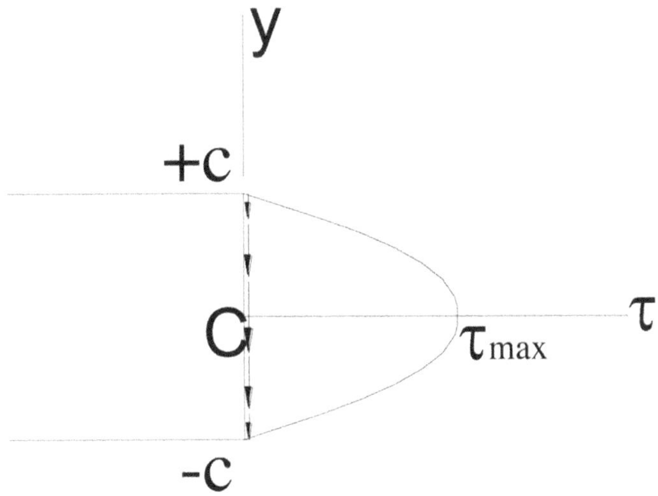

Whereas, steel is an expensive material in design. Additionally, small cross sections cannot be used for long compressive members because of slenderness. The protection of steel from environmental conditions are also needed. In that case, steel can be used with concrete to provide economy. Thus, reinforced concrete (RC) structures were developed.

The internal force diagrams of the frame given as Figure 3 and Figure 4 for two types of loading are shown as Figures 10 and 11 for distributed loads and lateral loads, respectively.

The distributed loads on structures are resulted from live and dead loads. Live loads are the ones which exist or not in a time. These loads cover humans and loads provided by humans, furniture and equipment, etc. Dead loads exist all time in structures and it also contains the self-weight of structural members.

Other load types result from earthquakes, winds, snow, traffic etc. which are applied to structures for a short time period. These loads show great differences according to region of construction. A place can expose strong winds, while another one is near to major active fault producing strong ground motions. For example, if the earthquake or wind loads are assumed as lateral loads, additionally by considering the change of the rotation of these loads, a place of members under compressive stress for distributed loading, while the same place can be under tensile stress during an earthquake. For example, at the connection points of beam, the flexural moment is negative. In that case, the upper site of beam must be reinforced with steel bars (rebar) for distributed load. For lateral loads, by considering direction changes of load,

Figure 10. The internal forces on frames under disturbed loads

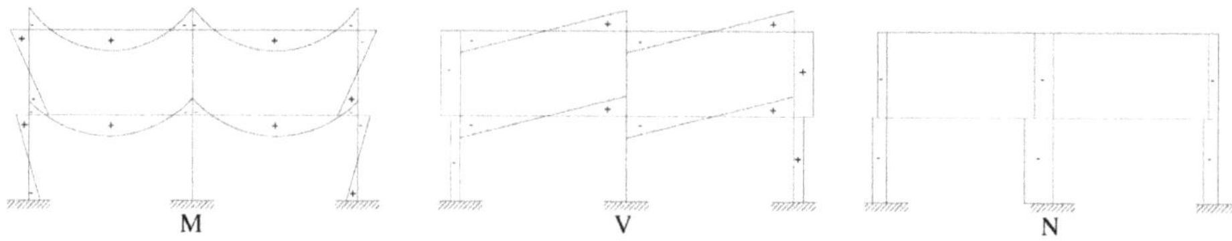

Figure 11. The internal forces on frames under disturbed loads

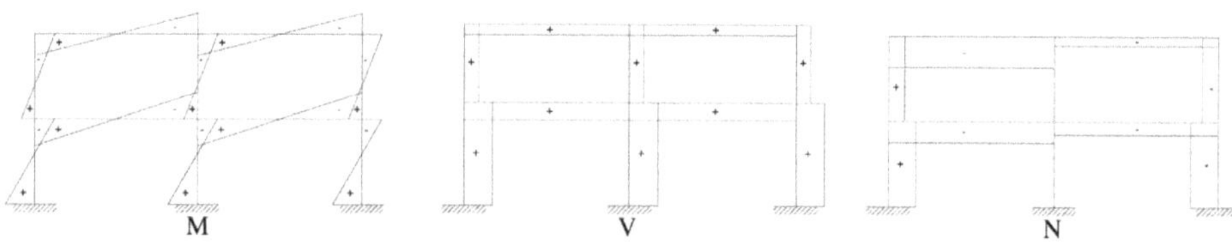

the moment may be negative to increase the intensity of the stress or it may be positive to change the superposed stress on the upper site. In the second case, the lower site will be under tensile stress and the lower part of the beam may need rebar. This situation makes the design and optimization process of RC structures complex.

Ductile behavior is also important for structures. Concrete is brittle material which is only effective on compressive stress and the yielding ratio is too low. In that case, the fracture of concrete is sudden. To ensure a ductile structural performance, steel bar plays and important role in additional to carry tensile stresses. For ductile behavior, the yielding of steel in beam member must start before the fracture of concrete. This situation cannot be provided for columns which have massive axial compressive forces. For that reason, several rules about ductile behavior are defined in design codes. The most basic rules in columns are to limit axial load respect to capacity of column cross section dimensions and providing frequent wrapping of longitudinal rebar. These conditions highly increase the number of design constraints and these constraints can be only checked if all dimensions and reinforcements are known. In that case, the optimization of RC member is highly nonlinear. It cannot be handled by mathematical methods. This reason is the fundamental use of metaheuristics in the optimization of RC structures.

BRIEF HISTORY OF OPTIMUM DESIGN OF STRUCTURES

In this section, the studies developed in the early period of structural optimization are given. Essentially, the oldest idea to optimize beam members were mentioned by Galilei Galileo in 1665 in order to improve the shape of statically determinate beams (Galilei, 1950). Early iterative computational efforts on optimization of beam were done by Hang and Krimser (1967) to minimize weight of beams using design constraints related to stress and deflection. A search procedure employing energy criteria was proposed by Venkaya (1971) to optimize structures including beams, frames and trusses by fully considering design constraints about sizes of elements, stress and displacement under multiple loading conditions.

For optimum design of RC beams, Friel (1974) developed optimum design formulation for ratio of steel to concrete area by considering singly RC beam subjected to moment constraint. By using Lagrange multipliers method, Chou (1977) optimized depth and area of flexural reinforcement of RC T-beam sections. Krishnamoorthy and Munro (1979) proposed a linear programming model for optimized RC frames with several types of constraints related to compatibility, limited ductility, equilibrium and serviceability. Kirsch (1983) proposed a three-level iterative methodology for cost optimization of multi-span continuous RC beam and reinforcements, cross-sectional dimensions and design moments are consequently handled. Lakshmanan and Parameswaran (1985) proposed a method for determination of span to effective depth ratios to avoid trial and error approaches of optimum flexural design of RC sections. Prakash, Agarwala and Singh (1988) developed simple optimization techniques for singly and doubly reinforced beams, T-beams and eccentricity loaded columns. Hoit, Soeiro, and Fagundo (1991) used augmented Lagrange multipliers and nonlinear programming techniques to reduce the weight of frames. Ghakrabaty (1992a, 1992b) developed an optimization model using several constraints by using geometric programing and Newton-Raphson methods. By taking the derivatives of augmented Lagrangian function, Al-Salloum and Siddiqi (1994) obtained a closed form solution for area and depth respect to cost and strength parameters. In the late 90s, computational methods including sequential linear programming and gradient projection method for RC beams (Chung & Sun 1994), continuum-type optimality criteria method for RC beams (Adamu, Karihaloo, & Rozvany, 1994), internal penalty function algorithm for RC short-tied columns (Zielinskil, Long, & Troitsky 1995), optimally criteria for 3d skeletal structures (Fadaee & Grierson 1994) and multilevel method for 3d frames (Balling & Yao 1997).

STATE OF ART REVIEW OF METAHEURISTIC-BASED OPTIMIZATION STUDIES OF REINFORCED CONCRETE STRUCTURES

In design of civil structures, the main purpose is generally the minimization of cost of engineering designs. The studies related to the optimum design of RC members or structures are mentioned in this section.

As all engineering optimization studies, genetic algorithm is also mostly used algorithm for RC design. Coello, Hernández and Farrera (1997) optimized RC beams with rectangular cross-sections by minimizing concrete, steel and formwork costs via GA. GA was also included in the optimization methodology of biaxial loaded RC columns supported by a declarative approach for capacity checking (Rafiq & Southcombe, 1998).

Koumousis and Arsenis (1998) employed GA to convert the required reinforcement area into a set of reinforcing bars described by specific diameter and length located at different parts of continuous beams. For RC structural members, Rath, Ahlawat, and Ramaswamy (1999) proposed a hybrid method using GA for cost optimization and sequential quadratic programming technique for shape optimization.

Camp, Pezeshk and Hansson (2003) employed GA based methodology to optimize basically supported beams, uniaxial loaded columns and multi-story frames.

Govindaraj and Ramasamy (2005) optimized RC continuous beams via GA based approach. In the approach, longitudinal reinforcements are considered as 4 groups and the possible combinations of diameters of the groups are chosen from a template including 861 combinations.

Simulated annealing (SA) which is also a classical metaheuristic method, have been included in RC optimization studies. Leps and Sejnoha (2003) proposed a hybrid method combining GA and SA for optimization of RC continuous beams.

Fedghouche and Tiliouine (2012) optimized RC beams with T-shaped cross-sections by using GA. In this approach, Eurocode 2 rules were considered as design constraints and these constrained were formulated for singly reinforced RC beam. These formulations are found in Chapter 3 and the optimization code using Jaya Algorithm (JA) is also presented.

Bekdaş and Nigdeli (2013) optimized T-shaped RC beam via an optimization approach employing Harmony Search and considering the detailed optimization of rebars in two lines of tensile and compressive sides of beam. HS was also employed for cost optimization of RC columns by considering slenderness by Bekdaş and Nigdeli (2014a, 2014b). Optimization of RC shear walls was also developed by using HS algorithm by Nigdeli and Bekdaş (2014a). The optimized design variables for RC shear walls are thickness of shear wall, web reinforcements of column headings, web reinforcements of shear wall and stirrups.

Nigdeli and Bekdaş (2017a) developed a random search technique (RST) for cost optimization of RC continuous beams. As seen in Fig. 13 for a two-span continuous beam, unfavorable live-load distributions were considered. In Fig. 13, LP defines load pattern and D and L are the dead and live distributed loads, respectively. For the solution of internal forces of continuous beam, three moment equations using Clapeyron's theory were used. The methodology employing RST uses three conditions to finalize the iteration of design variable generations. These three conditions are as follows:

1. Seven design constraints related to flexural capacity, shear capacity and ductile behavior must be provided.
2. For economy, randomly assigned reinforcements must exceed 5% more of required reinforcement for the critical cross-section.
3. If a doubly reinforced beam is needed, the reinforcements in the compression section must be less than the reinforcements in tensile section to prevent non-economical candidate solutions.

Jahjouh, Arafa and Alqedra (2013) optimize RC continuous beams by employing artificial bee colony (ABC) algorithm and modified ABC by including a variable changing percentage to improve the performance when dealing with members consisted of multiple variables

Akın and Saka (2010) employed HS for optimizing RC continuous beams by considering a detailed reinforcement positioning.

Bekdaş and Nigdeli (2015b) applied RST for RC columns subjected to axial load and flexural moment in one-dimension. The design variables of RC column were cross-sectional dimensions, bars in upper and lower section (in two lines), web reinforcements and shear reinforcements. Nigdeli, Bekdaş, Kim and Geem (2015) optimized biaxial loaded columns by using an improved HS method using different randomization stages in optimization process. As seen in Fig. 14, the column was under shear force (V_x), axial load (N_z), flexural moment around x axis (M_x) and y axis (M_y).

Figure 12. The optimized RC beam section
(Bekdaş & Nigdeli, 2018a)

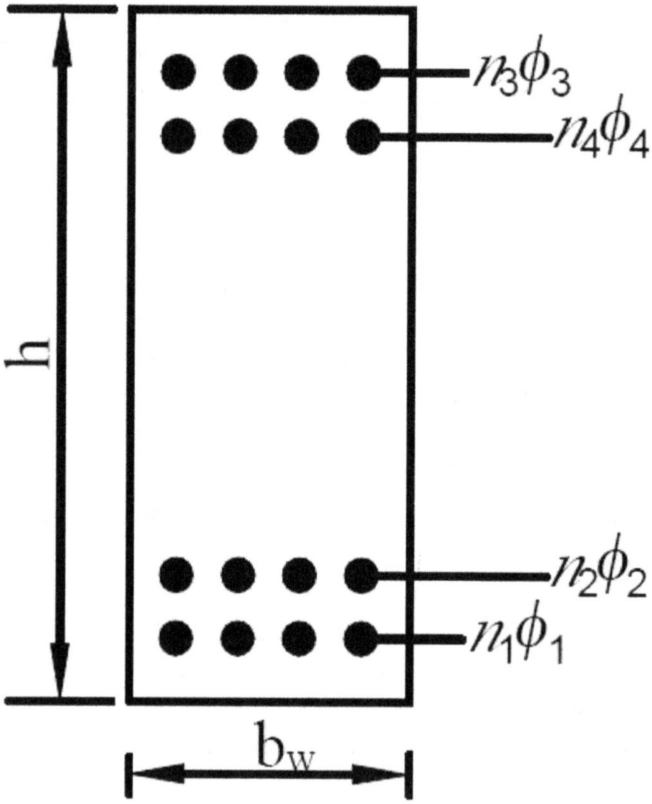

Figure 13. Loading condition for live-loads patterns of two-span continuous beam
(Nigdeli & Bekdaş, 2017a)

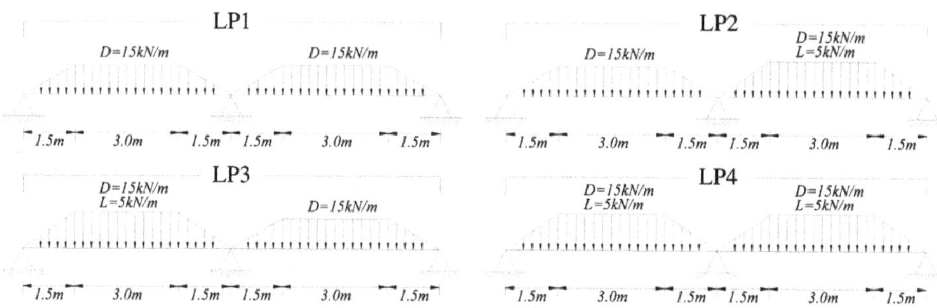

Figure 14. RC biaxial loaded column
(Nigdeli et al., 2015)

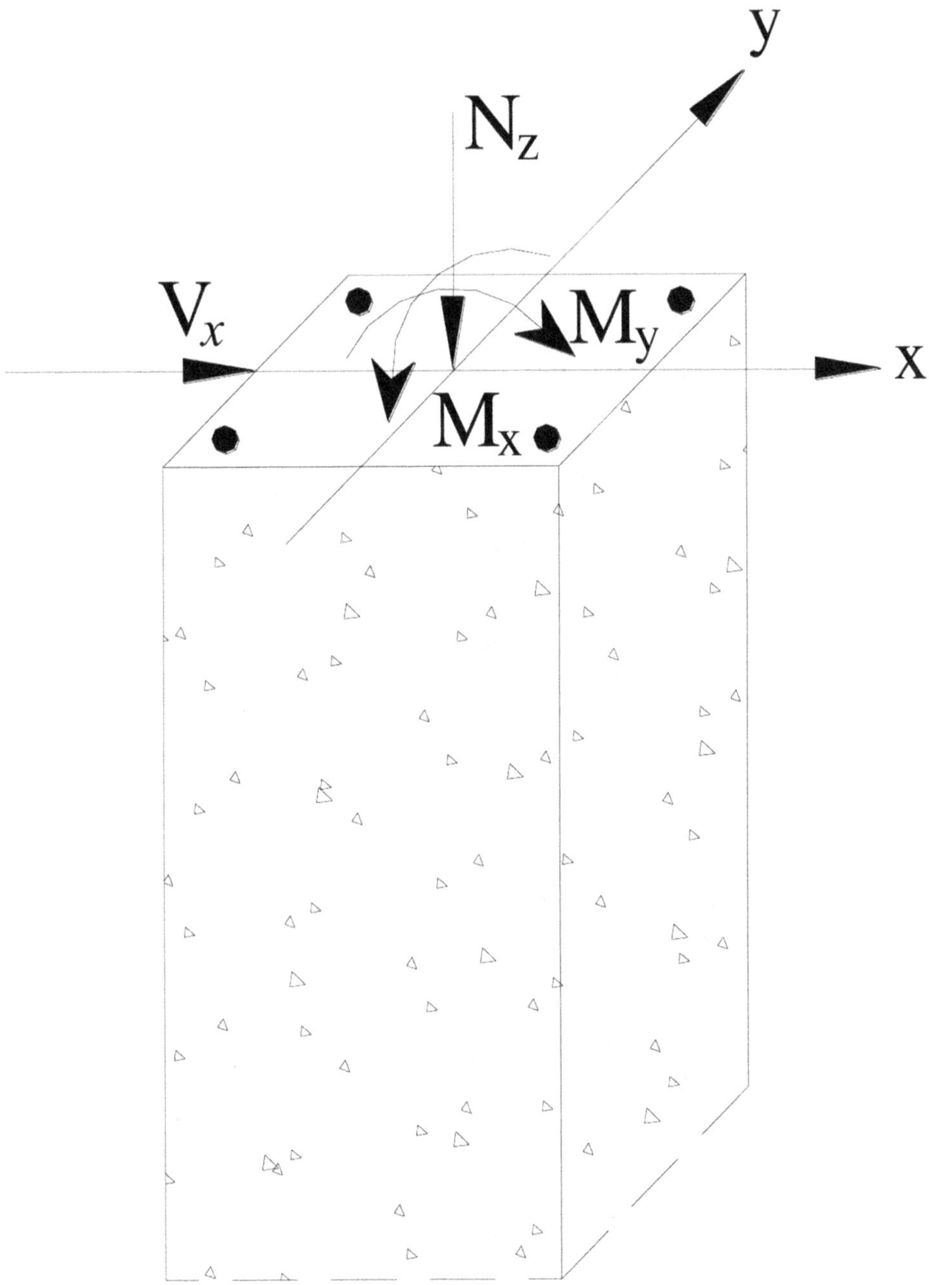

Nigdeli and Bekdaş (2014b) considered slenderness effects of RC column for cost optimization by using HS. A moment magnification factor considering second order effects of axial force of column is calculated in methodology according to ACI 318: Building code requirements for structural concrete and commentary. Bat algorithm (Bekdaş & Nigdeli, 2016a) and teaching learning-based optimization (Bekdaş & Nigdeli, 2016b) were also employed for slender RC columns.

GA and discrete form of Hook and Jeeves method are used as a hybrid method in optimum design of RC flat slab building (Sahab, Ashour, & Toropov, 2005a, 2005b). In this approach, GA is modified by Sahab et al. (2005a) to dynamically change the population size. The modified GA was used as a global optimization stage, while discretized form of Hook and Jeeves method was used for local optimization after global optimization stage to find better optimum results. Also, flat slab buildings are optimized in three stages including layout optimization of columns, dimension optimization of columns and slab for each column layout and reinforcement optimization with numbers and sizes by Sahab et al. (2005b). In addition to these, harmony search algorithm was employed by Kaveh and Abadi (2011) to minimize concrete and steel reinforcement costs of RC one-way joist floor system consisting of a hollow slab. Ghandi, Shokrollahi and Nasrolahi (2017) investigated optimum design with minimum cost of the one-way and two-way reinforced concrete (RC) slabs employing Cuckoo Optimization Algorithm (COA) and it was concluded that CAO is effective in finding optimum solutions.

Guerra and Kiousis (2006) optimized RC frames by using Sequential Quadratic Programming (SQP) for constrained nonlinear problem. Optimum sizing and reinforcing of beam and column members of multi-bay and multi-story frames were considered by implementing ACI 318 rules to SQP function of Maltab.

In road construction, cost optimization of RC box frames was investigated by the comparison of two heuristic algorithms such as random walk and the descent local search and two metaheuristic algorithms such as the threshold accepting and the simulated annealing. The threshold accepting is more efficient than others for frames of 13 m of horizontal span (Perea, Alcala, Yepes, Gonzalez-Vidosa, & Hospitaler, 2008).

Several well-known metaheuristic algorithms such as GA (Rajeev & Krishnamoorthy,1998; Camp et al., 2003; Lee & Ahn, 2003; Govindaraj & Ramasamy, 2007), SA (Paya-Zaforteza, Yepes, Hospitaler, & Gonzalez-Vidosa, 2008) and HS (Akin & Saka, 2015; Bekdaş & Nigdeli, 2014c; Ulusoy, Kayabekir, Bekdaş, & Nigdeli, 2018) has been employed in optimization of

RC frames for optimum cost. Also, hybrid methods were used for design optimization of RC frames for the similar purpose. To solve cost optimization problems, Kaveh and Sabzi (2011, 2012) employed Big Bang Big Crunch (BB-BC) algorithm and a hybrid algorithm combining Particle Swarm Optimization (PSO) and Ant Colony (ACO). Esfandiary, Sheikholarefin and Bondarabadi (2016) combined particle swarm optimization (PSO) and multi-criterion decision-making to minimize the material and construction cost of reinforced concrete. Chutani (2016) used particle swarm optimization (PSO) and standard gravitational search algorithm (GSA) together.

Generally, design is done under static load conditions. By including time history analysis of earthquake excitations, Nigdeli and Bekdaş developed a methodology using modified HS including several random search stages to skip unnecessary stages of different members of frames for different types of design constraints. This methodology was applied to multi-span and multi-story frames (Nigdeli & Bekdaş, 2016; Bekdaş & Nigdeli, 2017). Also, Arroyo and Gutiérrez (2017) aimed to improve the seismic performance of RC frame buildings and for this propose, a methodology called eigenfrequency optimization was utilized.

Differently from the cost minimization, CO_2 emission during the construction can be also minimized as done by Paya-Zoforteza et al. (2009) using SA and Camp and Huq (2013) using BB-BC. Martinez-Martin, Gonzalez-Vidosa, Hospitaler and Yepes (2012) developed several hybrid metaheuristic methods using SA and a neighborhood move based on the mutation operator of GA to solve multi-objective optimization problem of RC bridge piers. The considered objectives are cost minimization, reinforcing steel congestion and minimization of embedded CO_2 emissions. Park, Lee, Kim, Hong and Choi (2014) presented a parametric study considering structural materials in the structural design phase of RC columns for minimization of CO_2 emission or material cost. In the study conducted by Kaveh and Ardalani (2014), optimum design of frames providing both of minimization of CO_2 emission and material cost was investigated. In this multi-objective optimization process, Enhanced Colliding Bodies Optimization (ECBO) and the Non-dominated Sorting Enhanced Colliding Bodies Optimization (NSECBO) algorithms were used.

In addition to these methods, artificial intelligence methods have been used in optimization process. In this paragraph, some of these were summarized. The neural dynamics model developed by Adeli and Park (1995) was also employed in methodologies optimizing RC structures. Ahmadkhanlou and Adeli (2005) used the neural dynamics model to minimize cost of RC slabs according to ACI 318. Sirca Jr. and Adeli (2005) optimized pre-stressed concrete I-beam bridge system employing the neural dynamics model for cost minimization. In addition to these, neural network-based methods have been used for optimization of RC elements. Saini, Sehgal and Gambhir (2006) proposed genetically optimized artificial neural network (NN) based optimization for design of singly and doubly RC beams with minimum cost. Çağlar (2009) utilized Neural Network approach to determine the shear strength of circular reinforced concrete column. The artificial neural network model was employed for cost optimization of simply supported beams by Babiker, Adam and Mohamed (2012) considering the requirements of the ACI 318-08 code. Also, Kao and Yeh (2014) was used neural networks method in optimization of RC frames.

Civil engineering structures are constructed on soil. In that case, the property of soil is important, and structures are interacting with soil. Because of unknown characteristic of soil and complexity of the theory, general soil-structure interaction (SSI) is neglected or assumptions using big safety factors are done. For the major members and types of structures like footing and earth-retaining walls, geotechnical design constraints must be controlled. Optimization of footing and retaining walls is complex to other structures since both structural and geotechnical design constraints are handled together. This complexity urged optimum designers to use heuristic.

For earth retaining walls, geotechnical design constraints related to overturning stability, sliding stability and bearing capacity must be controlled by usage of different factor safety. Additionally, structural design constraints must be controlled in several parts of wall such as steam, toe, heel and shear key. Ceranic, Fryer and Baines (2001) proposed a SA-based methodology to minimize total material and construction costs. Yepes, Alcala, Perea and González-Vidosa (2008) investigated earth-retaining walls by developing a parametric study using optimum results found by using SA. Cantilever RC retaining walls were also optimized by employing HS to minimize cost (Kaveh & Abadi, 2011). Gahazavi and Salavati (2011) employed bacterial foraging optimization algorithm inspired by the social foraging behavior of Escherichia coli to solve cost minimization problem of RC cantilever retaining walls and a detailed sensitivity analysis of design variables, parameters and factor of safeties were included. Camp and Akin (2012) employed BB-BC for optimization of RC retaining walls for objectives of low-cost and low-weight by investigating example cases considering surcharge load, backfill slope, internal friction angle of the retained soil and effect of a base shear key. Talatahari, Sheikholeslami, Shadfaran and Pourbaba (2012) optimized gravity retaining walls by reducing total weight and charged system search (CSS) developed via the laws of electrostatic in physics were employed.

Kaveh, Kalateh-Ahani and Fahimi-Farzam (2013) optimized RC retaining walls by considering two objectives. In addition to cost minimization, reinforcing bar congestion was also taken as objective to provide constructability and the methodology employs non-dominated sorting genetic algorithm. Khajehzadeh, Taha and Eslami (2014) investigated the minimization of cost and CO_2 emission. For this multi objective optimization problem, a hybrid optimization method using together adaptive gravitational search algorithm (AGSA) with pattern search (PS) algorithms was proposed.

Gandomi, Kashani, Roke and Mousavi (2015) explored the efficiency of accelerated particle swarm optimization (APSO), firefly algorithm (FA), and cuckoo search (CS) on the finding optimum design of retaining walls and efficiency of these algorithms was demonstrated. The performance of two employed metaheuristic algorithms such as Colliding Particle Swarm Optimization was evaluated with cost optimization of RC retaining walls by Kaveh and Soleimani (2015) and the optimization was done according to earth pressures under static loading found according to Coulomb and Rankine theory and dynamic loading according to Mononobe-Okabe method.

Temür and Bekdaş (2016) optimized RC retaining walls under static loading by employing TLBO and improved TLBO proposed by Camp and Farschin (2014). TLBO was also employed for RC retaining wall optimization problem by considering dynamic loading (Kayabekir, Bekdaş, Nigdeli, & Temür, 2016) and space restriction of area of construction (Bekdaş, Nigdeli, Temür, & Kayabekir, 2016). Sheikholeslami, Khalili, Sadollah and Kim (2016) proposed also a hybrid method combining firefly algorithm and harmony search algorithm to find an economical RC design of retaining wall under statically loads. For the same purpose, Aydoğdu (2017) introduced biogeography-based optimization algorithm with Levy flight distribution (LFBBO) for the optimum design of retaining walls under the seismic loads. Also, harmony search algorithm with an intensification stage through threshold accepting was utilized for economical design of buttressed earth-retaining walls by Molina-Moreno (2017). Then, Grey wolf optimizer was used for optimum design of RC retaining wall including a shear key by Temür, Kayabekir, Bekdaş and Nigdeli (2018).

Footings are another structural member which must be designed according to both structural and geotechnical rules. Generally, RC spread footings were optimized by using metaheuristic methods. Khajehzadeh, Taha, El-Shafie and Eslami (2011) employed a modified PSO for optimum design of RC spread footing and retaining walls. The optimized spread footing has a column loaded with biaxial flexural moment. Gravitational Search algorithm developed according to Newton's Law is another heuristic method employed for single spread footings (Khajehzadeh, Taha, El-Shafie, & Eslami, 2012). As a second objective, CO_2 emission minimization was also considered in addition to cost minimization of spread footing in studies employing metaheuristic such as a hybrid algorithm integrating the firefly algorithm and sequential quadratic programming (Khajehzadeh, Taha, & Eslami, 2013) and hybrid of BB-BC algorithm (Camp & Assadollahi, 2013). For RC spread footings, HS was employed to find the dimension and reinforcement design variables by considering both structural design and geotechnical design constraints such as bearing pressure and settlement of footing (Bekdaş & Nigdeli, 2015c). PSO was also employed for the same methodology (Nigdeli & Bekdaş, 2017b). The multi-stage methodology considering dimension and reinforcement values consequently by using separate stages was also investigated by employing Flower Pollination algorithm (FPA) and teaching-learning-based optimization (TLBO). The dimension design variables of RC footing were detailly optimized by considering shape of footing and eccentricity of the mounted column (Nigdeli & Bekdaş, 2018).

Reinforced concrete structures can be retrofitted to increase capacity respect to internal forces. A non-destructive method of retrofit is to cover RC section with carbon fiber reinforced polymer (CFRP) strips. To increase shear force capacity of RC beams, the optimum width, spacing and angle of CFRP were investigated by using metaheuristic methods such as FPA, TLBO (Kayabekir, Sayin, Bekdaş, & Nigdeli, 2018b) and Jaya algorithm (Kayabekir, Bekdaş, Nigdeli, & Temür, 2018a; Kayabekir, Sayın, Nigdeli, & Bekdaş, 2018c).

The optimum design providing minimum cost of reinforced concrete vaults used in road construction was investigated by Carbonell, González-Vidosa and Yepes (2011). In the optimization process, the multi-start global best descent local search (MGB), the meta-simulated annealing (SA) and the meta-threshold acceptance (TA) methods are employed and it was concluded that the MGB is more efficient in finding the optimum results for RC vaults.

RC Liquid tanks modelled as axially symmetric cylindrical wall were optimized for thickness and compressive strength of concrete by using HS with the superposition method for cost minimization (Bekdaş 2014). In addition to these design variables, locations, intensities of the post tensioned loads and the diameter of the rebars and distance were considered as design variables, too (Bekdaş 2015). Differently from cost minimization, the minimization of maximum longitudinal moment on the wall was provided by the optimum placement of post-tensioning cables providing optimum forces. For this application, four different metaheuristic methods such as bat algorithm, FPA, TLBO and HS were compared (Bekdaş & Nigdeli, 2018b) the optimization of thickness by checking structural constraints and both considering optimum post-tensioning forces with the cost of cables with application was also investigated by using HS-based methodology (Bekdaş, 2018). Then, the same optimization was done by using hybrid methods to increase the efficiency of optimum design (Bekdaş, 2019).

REFERENCES

ACI-318. (2005). Building code requirements for structural concrete and commentary, metric version. *American Concrete Institute.*

Adamu, A., Karihaloo, B. L., & Rozvany, G. I. N. (1994). Minimum cost design of reinforced concrete beams using continuum-type optimality criteria. *Structural Optimization*, *7*(1-2), 91–102. doi:10.1007/BF01742512

Adeli, H., & Park, H. S. (1995). Optimization of space structures by neural dynamics. *Neural Networks*, *8*(5), 769–781. doi:10.1016/0893-6080(95)00026-V

Ahmadkhanlou, F., & Adeli, H. (2005). Optimum cost design of reinforced concrete slabs using neural dynamics model. *Engineering Applications of Artificial Intelligence*, *18*(1), 65–72. doi:10.1016/j.engappai.2004.08.025

Akin, A., & Saka, M. P. (2010). Optimum Detailed Design of Reinforced Concrete Continuous Beams using the Harmony Search Algorithm, In *The Tenth International Conference on Computational Structures Technology*, Paper 131, Stirlingshire, UK.

Akin, A., & Saka, M. P. (2015). Harmony search algorithm based optimum detailed design of reinforced concrete plane frames subject to ACI 318-05 provisions. *Computers & Structures*, *147*, 79–95. doi:10.1016/j.compstruc.2014.10.003

Al-Salloum, Y. A., & Husainsiddiqi, G. (1994). Cost-optimum design of reinforced concrete (RC) beams. *Structural Journal*, *91*(6), 647–655.

Arroyo, O., & Gutiérrez, S. (2017). A seismic optimization procedure for reinforced concrete framed buildings based on eigenfrequency optimization. *Engineering Optimization*, *49*(7), 1166–1182. doi:10.1080/0305215X.2016.1241779

Aydogdu, I. (2017). Cost optimization of reinforced concrete cantilever retaining walls under seismic loading using a biogeography-based optimization algorithm with Levy flights. *Engineering Optimization*, *49*(3), 381–400. doi:10.1080/0305215X.2016.1191837

Babiker, S., Adam, F., & Mohamed, A. (2012). Design optimization of reinforced concrete beams using artificial neural network. *Int. J. Eng. Inventions, 1*(8), 7-13.

Balling, R. J., & Yao, X. (1997). Optimization of reinforced concrete frames. *Journal of Structural Engineering*, *123*(2), 193–202. doi:10.1061/(ASCE)0733-9445(1997)123:2(193)

Bekdas, G. (2014). Optimum design of axially symmetric cylindrical reinforced concrete walls. *Structural Engineering and Mechanics*, *51*(3), 361–375. doi:10.12989em.2014.51.3.361

Bekdaş, G. (2015). Harmony search algorithm approach for optimum design of post-tensioned axially symmetric cylindrical reinforced concrete walls. *Journal of Optimization Theory and Applications*, *164*(1), 342–358. doi:10.100710957-014-0562-2

Bekdaş, G. (2018). New improved metaheuristic approaches for optimum design of posttensioned axially symmetric cylindrical reinforced concrete walls. *Structural Design of Tall and Special Buildings*, *27*(7). doi:10.1002/tal.1461

Bekdaş, G. (2019). Optimum design of post-tensioned axially symmetric cylindrical walls using novel hybrid metaheuristic methods. *Structural Design of Tall and Special Buildings*, *28*(1). doi:10.1002/tal.1550

Bekdaş, G., & Nigdeli, S. M. (2013). Optimization of T-shaped RC flexural members for different compressive strengths of concrete. *International Journal of Mechanics*, *7*, 109–119.

Bekdas, G., & Nigdeli, S. M. (2014a). The optimization of slender reinforced concrete columns. *Proceedings in Applied Mathematics and Mechanics*, *14*(1), 183–184. doi:10.1002/pamm.201410079

Bekdaş, G., & Nigdeli, S. M. (2014b). The Effect of Eccentricity for Optimum Compressively Loaded Reinforced Concrete Column. In *15th EU/ME Workshop: Metaheuristic and Engineering*, Istanbul, Turkey. Academic Press.

Bekdaş, G., & Nigdeli, S. M. (2014c) Optimization of RC Frame Structures Subjected to Static Loading. In *11th World Congress on Computational Mechanics*, Barcelona, Spain. Academic Press.

Bekdaş, G., & Nigdeli, S. M. (2015a). Optimum Design of Reinforced Concrete Beams Using Teaching-Learning-Based Optimization. In *3rd International Conference on Optimization Techniques in Engineering (OTENG '15)*, Rome, Italy. Academic Press.

Bekdaş, G., & Nigdeli, S. M. (2015b). Optimization of Reinforced Concrete Columns Subjected to Uniaxial Loading. In N. Lagaros, & M. Papadrakakis (Eds.), *Engineering and Applied Sciences Optimization*. Springer. doi:10.1007/978-3-319-18320-6_21

Bekdaş, G., & Nigdeli, S. M. (2015c). Multi-objective optimization of reinforced concrete footings using harmony search. In *23rd International Conference on Multiple Criteria Decision Making (MCDM 2015)*, Hamburg, Germany. Academic Press.

Bekdaş, G., & Nigdeli, S. M. (2016a). T Bat algorithm for optimization of reinforced concrete columns. In *Joint Annual Meeting of GAMM and DMV*, Braunschweig, Germany. 10.1002/pamm.201610329

Bekdaş, G., & Nigdeli, S. M. (2016b). Optimum Design Of Reinforced Concrete Columns Employing Teaching Learning Based Optimization. In *12th International Congress on Advances in Civil Engineering*, Istanbul, Turkey. Academic Press.

Bekdaş, G., & Nigdeli, S. M. (2017). Modified Harmony Search for Optimization of Reinforced Concrete Frames. In *3rd International Conference on the Harmony Search Algorithm (ICHSA 2017)*, Bilbao, Spain. 10.1007/978-981-10-3728-3_21

Bekdaş, G., & Nigdeli, S. M. (2018a). Robustness of Metaheuristic Algorithms in Optimum Design of Reinforced Concrete Beams. In *International Conference on Bioinspired Optimization Methods and their Applications (BIOMA 2018)*, Paris, France. Academic Press.

Bekdaş, G., & Nigdeli, S. M. (2018b). Optimum reduction of flexural effect of axially symmetric cylindrical walls with post-tensioning forces. *KSCE Journal of Civil Engineering*, 1–8.

Bekdas, G., Nigdeli, S. M., & Kayabekir, A. E. (2017). Optimum Design of Reinforced Concrete Shear Walls employing Teaching Learning Based Optimization. In *2nd International Conference on Civil and Environmental Engineering (ICOCEE 2017)*, Cappadocia, Turkey. Academic Press.

Bekdas, G., Nigdeli, S. M., Temür, R., & Kayabekir, A. E. (2016). Restricted Optimum Design of Reinforced Concrete Retaining Walls. In *7th European Conference of Civil Engineering (ECCIE '16)*. Bern, Switzerland. Academic Press.

Bekdaş, G., Nigdeli, S. M., & Yang, X.-S. (2014). Metaheuristic Optimization for the Design of Reinforced Concrete Beams under Flexure Moments. In *5th European Conference of Civil Engineering (ECCIE '14)*. Florence, Italy. Academic Press.

Bekdaş, G., Nigdeli, S. M., & Yang, X. S. (2015). Sizing optimization of truss structures using flower pollination algorithm. *Applied Soft Computing*, *37*, 322–331. doi:10.1016/j.asoc.2015.08.037

Caglar, N. (2009). Neural network-based approach for determining the shear strength of circular reinforced concrete columns. *Construction & Building Materials*, *23*(10), 3225–3232. doi:10.1016/j.conbuildmat.2009.06.002

Camp, C. V. (2007). Design of space trusses using Big Bang–Big Crunch optimization. *Journal of Structural Engineering*, *133*(7), 999–1008. doi:10.1061/(ASCE)0733-9445(2007)133:7(999)

Camp, C. V., & Akin, A. (2011). Design of retaining walls using big bang–big crunch optimization. *Journal of Structural Engineering*, *138*(3), 438–448. doi:10.1061/(ASCE)ST.1943-541X.0000461

Camp, C. V., & Assadollahi, A. (2013). CO2 and cost optimization of reinforced concrete footings using a hybrid big bang-big crunch algorithm. *Structural and Multidisciplinary Optimization*, *48*(2), 411–426. doi:10.100700158-013-0897-6

Camp, C. V., & Bichon, B. J. (2004). Design of space trusses using ant colony optimization. *Journal of Structural Engineering*, *130*(5), 741–751. doi:10.1061/(ASCE)0733-9445(2004)130:5(741)

Camp, C. V., Bichon, B. J., & Stovall, S. P. (2005). Design of steel frames using ant colony optimization. *Journal of Structural Engineering*, *131*(3), 369–379. doi:10.1061/(ASCE)0733-9445(2005)131:3(369)

Camp, C. V., & Farshchin, M. (2014). Design of space trusses using modified teaching–learning based optimization. *Engineering Structures*, *62*, 87–97. doi:10.1016/j.engstruct.2014.01.020

Camp, C. V., & Huq, F. (2013). CO2 and cost optimization of reinforced concrete frames using a big bang-big crunch algorithm. *Engineering Structures*, *48*, 363–372. doi:10.1016/j.engstruct.2012.09.004

Camp, C. V., Pezeshk, S., & Hansson, H. (2003). Flexural design of reinforced concrete frames using a genetic algorithm. *Journal of Structural Engineering*, *129*(1), 105–115. doi:10.1061/(ASCE)0733-9445(2003)129:1(105)

Carbonell, A., González-Vidosa, F., & Yepes, V. (2011). Design of reinforced concrete road vaults by heuristic optimization. *Advances in Engineering Software*, *42*(4), 151–159. doi:10.1016/j.advengsoft.2011.01.002

Ceranic, B., Fryer, C., & Baines, R. W. (2001). An application of simulated annealing to the optimum design of reinforced concrete retaining structures. *Computers & Structures*, *79*(17), 1569–1581. doi:10.1016/S0045-7949(01)00037-2

Chakrabarty, B. K. (1992). Models for optimal design of reinforced concrete beams. *Computers & Structures*, *42*(3), 447–451. doi:10.1016/0045-7949(92)90040-7

Chakrabarty, B. K. (1992). Model for Optimal Design of Reinforced Concrete Beam. *Journal of Structural Engineering*, *118*(11), 3238–3242. doi:10.1061/(ASCE)0733-9445(1992)118:11(3238)

Chou, T. (1977). Optimum reinforced concrete T-beam sections. *Journal of the Structural Division*, 103(ASCE 13120).

Chung, T. T., & Sun, T. C. (1994). Weight optimization for flexural reinforced concrete beams with static nonlinear response. *Structural Optimization*, *8*(2-3), 174–180. doi:10.1007/BF01743315

Chutani, S., & Singh, J. (2018). Use of modified hybrid PSOGSA for optimum design of RC frame. *Zhongguo Gongcheng Xuekan*, *41*(4), 342–352. doi:10.1080/02533839.2018.1473804

Coello, C. A., & Christiansen, A. D. (2000). Multiobjective optimization of trusses using genetic algorithms. *Computers & Structures, 75*(6), 647–660. doi:10.1016/S0045-7949(99)00110-8

Coello, C. C., Hernández, F. S., & Farrera, F. A. (1997). Optimal design of reinforced concrete beams using genetic algorithms. *Expert Systems with Applications, 12*(1), 101–108. doi:10.1016/S0957-4174(96)00084-X

Dede, T., & Ayvaz, Y. (2015). Combined size and shape optimization of structures with a new meta-heuristic algorithm. *Applied Soft Computing, 28*, 250–258. doi:10.1016/j.asoc.2014.12.007

Degertekin, S. O. (2008). Optimum design of steel frames using harmony search algorithm. *Structural and Multidisciplinary Optimization, 36*(4), 393–401. doi:10.100700158-007-0177-4

Degertekin, S. O., & Hayalioglu, M. S. (2013). Sizing truss structures using teaching-learning-based optimization. *Computers & Structures, 119*, 177–188. doi:10.1016/j.compstruc.2012.12.011

Esfandiary, M. J., Sheikholarefin, S., & Bondarabadi, H. R. (2016). A combination of particle swarm optimization and multi-criterion decision-making for optimum design of reinforced concrete frames. *International Journal of Optimization in Civil Engineering, 6*(2), 245–268.

European Committee for Standardization. (2002, July). Eurocode 2: Design of Concrete Structures. Part 1: General rules and rules for buildings. Final Draft.

Fadaee, M. J., & Grierson, D. E. (1996). Design optimization of 3D reinforced concrete structures. *Structural Optimization, 12*(2-3), 127–134. doi:10.1007/BF01196945

Fedghouche, F., & Tiliouine, B. (2012). Minimum cost design of reinforced concrete T-beams at ultimate loads using Eurocode2. *Engineering Structures, 42*, 43–50. doi:10.1016/j.engstruct.2012.04.008

Friel, L. L. (1974, November). Optimum singly reinforced concrete sections. In Journal Proceedings, 71(11), 556-558.

Galilei, G. (1950). *Dialogues concerning two new sciences*. Evanston, IL: Northwestern University Press. (Original work published 1665)

Gandomi, A. H., Kashani, A. R., Roke, D. A., & Mousavi, M. (2015). Optimization of retaining wall design using recent swarm intelligence techniques. *Engineering Structures*, *103*, 72–84. doi:10.1016/j.engstruct.2015.08.034

Gandomi, A. H., Talatahari, S., Yang, X. S., & Deb, S. (2013). Design optimization of truss structures using cuckoo search algorithm. *Structural Design of Tall and Special Buildings*, *22*(17), 1330–1349. doi:10.1002/tal.1033

Ghandi, E., Shokrollahi, N., & Nasrolahi, M. (2017). Optimum cost design of reinforced concrete slabs using cuckoo search optimization algorithm. Iran University of Science & Technology, 7(4), 539-564.

Ghazavi, M., & Salavati, V. (2011). Sensitivity analysis and design of reinforced concrete cantilever retaining walls using bacterial foraging optimization algorithm. In *Proceedings of the 3rd International Symposium on Geotechnical Safety and Risk (ISGSR), München* (pp. 307-314).

Govindaraj, V., & Ramasamy, J. V. (2005). Optimum detailed design of reinforced concrete continuous beams using genetic algorithms. *Computers & Structures*, *84*(1-2), 34–48. doi:10.1016/j.compstruc.2005.09.001

Govindaraj, V., & Ramasamy, J. V. (2007). Optimum detailed design of reinforced concrete frames using genetic algorithms. *Engineering Optimization*, *39*(4), 471–494. doi:10.1080/03052150601180767

Guerra, A., & Kiousis, P. D. (2006). Design optimization of reinforced concrete structures. *Computers and Concrete*, *3*(5), 313–334. doi:10.12989/cac.2006.3.5.313

Hasançebi, O., Çarbaş, S., Doğan, E., Erdal, F., & Saka, M. P. (2010). Comparison of non-deterministic search techniques in the optimum design of real size steel frames. *Computers & Structures*, *88*(17-18), 1033–1048. doi:10.1016/j.compstruc.2010.06.006

Hasançebi, O., & Kazemzadeh Azad, S. (2014). Discrete size optimization of steel trusses using a refined big bang–big crunch algorithm. *Engineering Optimization*, *46*(1), 61–83. doi:10.1080/0305215X.2012.748047

Haug, E. J. Jr, & Kirmser, P. G. (1967). Minimum weight design of beams with inequality constraints on stress and deflection. *Journal of Applied Mechanics*, *34*(4), 999–1004. doi:10.1115/1.3607869

Hoit, M., Soeiro, A., & Fagundo, F. (1991). Probabilistic design and optimization of reinforced concrete frames. *Engineering Optimization*, *17*(3), 229–235. doi:10.1080/03052159108941072

Jahjouh, M. M., Arafa, M. H., & Alqedra, M. A. (2013). Artificial Bee Colony (ABC) algorithm in the design optimization of RC continuous beams. *Structural and Multidisciplinary Optimization*, *47*(6), 963–979. doi:10.100700158-013-0884-y

Kao, C. S., & Yeh, I. (2014). Optimal design of reinforced concrete plane frames using artificial neural networks. *Computers and Concrete*, *14*(4), 445–462. doi:10.12989/cac.2014.14.4.445

Kaveh, A., & Abadi, A. S. M. (2011). Harmony search-based algorithms for the optimum cost design of reinforced concrete cantilever retaining walls. *International Journal of Civil Engineering*, *9*(1), 1–8.

Kaveh, A., & Abadi, A. S. M. (2011). Cost optimization of reinforced concrete one-way ribbed slabs using harmony search algorithm. *Arabian Journal for Science and Engineering*, *36*(7), 1179–1187. doi:10.100713369-011-0113-1

Kaveh, A., & Ardalani, S. (2016). Cost and CO2 emission optimization of reinforced concrete frames using enhanced colliding bodies algorithm. *Asian Journal of Civil Engineering*, *17*(6), 831–858.

Kaveh, A., & Bakhshpoori, T. (2013). Optimum design of steel frames using Cuckoo Search algorithm with Lévy flights. *Structural Design of Tall and Special Buildings*, *22*(13), 1023–1036. doi:10.1002/tal.754

Kaveh, A., Kalateh-Ahani, M., & Fahimi-Farzam, M. (2013). Constructability optimal design of reinforced concrete retaining walls using a multi-objective genetic algorithm. *Structural Engineering and Mechanics*, *47*(2), 227–245. doi:10.12989em.2013.47.2.227

Kaveh, A., & Sabzi, O. (2011). A comparative study of two meta-heuristic algorithms for optimum design of reinforced concrete frames. *International Journal of Civil Engineering*, *9*(3), 193–206.

Kaveh, A., & Sabzi, O. (2012). Optimal design of reinforced concrete frames using big bang-big crunch algorithm. *International Journal of Civil Engineering*, *10*(3), 189–200.

Kaveh, A., & Soleimani, N. (2015). CBO and DPSO for optimum design of reinforced concrete cantilever retaining walls. *Asian J Civil Eng*, *16*(6), 751–774.

Kaveh, A., & Talatahari, S. (2009a). Particle swarm optimizer, ant colony strategy and harmony search scheme hybridized for optimization of truss structures. *Computers & Structures*, *87*(5-6), 267–283. doi:10.1016/j.compstruc.2009.01.003

Kaveh, A., & Talatahari, S. (2009b). Size optimization of space trusses using Big Bang–Big Crunch algorithm. *Computers & Structures*, *87*(17-18), 1129–1140. doi:10.1016/j.compstruc.2009.04.011

Kayabekir, A. E., Bekdaş, G., Nigdeli, S. M., & Temür, R. (2016). Optimum design of retaining walls under static and dynamic loads, In *International Science Symposium (ISS2016)*, Istanbul, Turkey. Academic Press.

Kayabekir, A. E., Bekdaş, G., Nigdeli, S. M., & Temür, R. (2018a). Investigation of cross-sectional dimension on optimum carbon fiber reinforced polymer design for shear capacity increase of reinforced concrete beams. In *7th International Conference on Applied and Computational Mathematics (ICACM '18)*. Rome, Italy. Academic Press

Kayabekir, A. E., Sayin, B., Bekdas, G., & Nigdeli, S. M. (2018b). The factor of optimum angle of carbon fiber reinforced polymers. In *4th International Conference on Engineering and Natural Sciences (ICENS 2018)*. Kiev, Ukraine. Academic Press.

Kayabekir, A. E., Sayın, B., Nigdeli, S. M., & Bekdaş, G. (2018c). Jaya algorithm based optimum carbon fiber reinforced polymer design for reinforced concrete beams. In AIP Conference Proceedings (vol. 1978, No. 1). AIP Publishing. doi:10.1063/1.5043891

Kelesoglu, O. (2007). Fuzzy multiobjective optimization of truss-structures using genetic algorithm. *Advances in Engineering Software*, *38*(10), 717–721. doi:10.1016/j.advengsoft.2007.03.003

Khajehzadeh, M., Taha, M. R., El-Shafie, A., & Eslami, M. (2011). Modified particle swarm optimization for optimum design of spread footing and retaining wall. *Journal of Zhejiang University. Science A*, *12*(6), 415–427. doi:10.1631/jzus.A1000252

Khajehzadeh, M., Taha, M. R., El-shafie, A., & Eslami, M. (2012). Optimization of Shallow Foundation Using Gravitational Search Algorithm. *Journal of Applied Engineering and Technology*, *4*(9), 1124–1130.

Khajehzadeh, M., Taha, M. R., & Eslami, M. (2013). A new hybrid firefly algorithm for foundation optimization. *National Academy Science Letters*, *36*(3), 279–288. doi:10.100740009-013-0129-z

Khajehzadeh, M., Taha, M. R., & Eslami, M. (2014). Multi-objective optimisation of retaining walls using hybrid adaptive gravitational search algorithm. *Civil Engineering and Environmental Systems*, *31*(3), 229–242. doi:10.1080/10286608.2013.853746

Kirsch, U. (1983). Multilevel optimal design of reinforced concrete structures. *Engineering Optimization*, *6*(4), 207–212. doi:10.1080/03052158308902471

Koumousis, V. K., & Arsenis, S. J. (1998). Genetic algorithms in optimal detailed design of reinforced concrete members. *Computer-Aided Civil and Infrastructure Engineering*, *13*(1), 43–52. doi:10.1111/0885-9507.00084

Koumousis, V. K., & Georgiou, P. G. (1994). Genetic algorithms in discrete optimization of steel truss roofs. *Journal of Computing in Civil Engineering*, *8*(3), 309–325. doi:10.1061/(ASCE)0887-3801(1994)8:3(309)

Krishnamoorthy, C. S., & Munro, J. (1973). Linear program for optimal design of reinforced concrete frames. *Proceedings of IABSE*, *3*(1), 119–141.

Lakshmanan, N., & Parameswaran, V. S. (1985). Minimum weight design of reinforced concrete sections for flexure. Journal of the Institution of Engineers. India. Civil Engineering Division, 66(2), 92-98.

Lee, C., & Ahn, J. (2003). Flexural design of reinforced concrete frames by genetic algorithm. *Journal of Structural Engineering*, *129*(6), 762–774. doi:10.1061/(ASCE)0733-9445(2003)129:6(762)

Lepš, M., & Šejnoha, M. (2003). New approach to optimization of reinforced concrete beams. *Computers & Structures*, *81*(18-19), 1957–1966. doi:10.1016/S0045-7949(03)00215-3

Li, J. P. (2015). Truss topology optimization using an improved species-conserving genetic algorithm. *Engineering Optimization*, *47*(1), 107–128. doi:10.1080/0305215X.2013.875165

Li, L. J., Huang, Z. B., Liu, F., & Wu, Q. H. (2007). A heuristic particle swarm optimizer for optimization of pin connected structures. *Computers & Structures*, *85*(7-8), 340–349. doi:10.1016/j.compstruc.2006.11.020

Martinez-Martin, F. J., Gonzalez-Vidosa, F., Hospitaler, A., & Yepes, V. (2012). Multi-objective optimization design of bridge piers with hybrid heuristic algorithms. *Journal of Zhejiang University. Science A*, *13*(6), 420–432. doi:10.1631/jzus.A1100304

Mathworks. (2010). MATLAB R2010a. The MathWorks Inc., Natick, MA.

Miguel, L. F. F., Lopez, R. H., & Miguel, L. F. F. (2013). Multimodal size, shape, and topology optimisation of truss structures using the Firefly algorithm. *Advances in Engineering Software*, *56*, 23–37. doi:10.1016/j.advengsoft.2012.11.006

Molina-Moreno, F., García-Segura, T., Martí, J. V., & Yepes, V. (2017). Optimization of buttressed earth-retaining walls using hybrid harmony search algorithms. *Engineering Structures*, *134*, 205–216. doi:10.1016/j.engstruct.2016.12.042

Nigdeli, S. M., & Bekdaş, G. (2014a). Optimization of reinforced concrete shear walls using harmony search. In *11th International Congress on Advances in Civil Engineering*, Istanbul, Turkey. Academic Press.

Nigdeli, S. M., & Bekdaş, G. (2014b). Optimum Design of RC Columns According to Effective Length Factor in Buckling. In *The Twelfth International Conference on Computational Structures Technology*, Naples, Italy. 10.4203/ccp.106.231

Nigdeli, S. M., & Bekdas, G. (2016). Detailed Optimum Design of Reinforced Concrete Frame Structures. In *7th European Conference of Civil Engineering (ECCIE '16)*, Bern, Switzerland. Academic Press.

Nigdeli, S. M., & Bekdaş, G. (2017a). Optimum design of RC continuous beams considering unfavourable live-load distributions. *KSCE Journal of Civil Engineering*, *21*(4), 1410–1416. doi:10.100712205-016-2045-5

Nigdeli, S. M., & Bekdas, G. (2017b) Optimization of Reinforced Concrete Footings Using Particle Swarm Optimization. 2nd International Conference on Civil and Environmental Engineering (ICOCEE 2017), Cappadocia, Turkey. Academic Press.

Nigdeli, S. M., Bekdas, G., Kim, S., & Geem, Z. W. (2015). A novel harmony search-based optimization of reinforced concrete biaxially loaded columns. *Structural Engineering and Mechanics*, *54*(6), 1097–1109. doi:10.12989em.2015.54.6.1097

Nigdeli, S. M., Bekdaş, G., & Yang, X. S. (2018). Metaheuristic optimization of reinforced concrete footings. *KSCE Journal of Civil Engineering*, *22*(11), 4555–4563. doi:10.100712205-018-2010-6

Park, H. S., Lee, H., Kim, Y., Hong, T., & Choi, S. W. (2014). Evaluation of the influence of design factors on the CO2 emissions and costs of reinforced concrete columns. *Energy and Building*, *82*, 378–384. doi:10.1016/j.enbuild.2014.07.038

Paya, I., Yepes, V., González-Vidosa, F., & Hospitaler, A. (2008). Multiobjective optimization of concrete frames by simulated annealing. *Computer-Aided Civil and Infrastructure Engineering*, *23*(8), 596–610. doi:10.1111/j.1467-8667.2008.00561.x

Paya-Zaforteza, I., Yepes, V., Hospitaler, A., & Gonzalez-Vidosa, F. (2009). CO2-optimization of reinforced concrete frames by simulated annealing. *Engineering Structures*, *31*(7), 1501–1508. doi:10.1016/j.engstruct.2009.02.034

Perea, C., Alcala, J., Yepes, V., Gonzalez-Vidosa, F., & Hospitaler, A. (2008). Design of reinforced concrete bridge frames by heuristic optimization. *Advances in Engineering Software*, *39*(8), 676–688. doi:10.1016/j.advengsoft.2007.07.007

Perez, R. L., & Behdinan, K. (2007). Particle swarm approach for structural design optimization. *Computers & Structures*, *85*(19-20), 1579–1588. doi:10.1016/j.compstruc.2006.10.013

Prakash, A., Agarwala, S. K., & Singh, K. K. (1988). Optimum design of reinforced concrete sections. *Computers & Structures*, *30*(4), 1009–1011. doi:10.1016/0045-7949(88)90142-3

Rafiq, M. Y., & Southcombe, C. (1998). Genetic algorithms in optimal design and detailing of reinforced concrete biaxial columns supported by a declarative approach for capacity checking. *Computers & Structures*, *69*(4), 443–457. doi:10.1016/S0045-7949(98)00108-4

Rajeev, S., & Krishnamoorthy, C. S. (1992). Discrete optimization of structures using genetic algorithms. *Journal of Structural Engineering*, *118*(5), 1233–1250. doi:10.1061/(ASCE)0733-9445(1992)118:5(1233)

Rajeev, S., & Krishnamoorthy, C. S. (1998). Genetic algorithm–based methodology for design optimization of reinforced concrete frames. *Computer-Aided Civil and Infrastructure Engineering*, *13*(1), 63–74. doi:10.1111/0885-9507.00086

Rath, D. P., Ahlawat, A. S., & Ramaswamy, A. (1999). Shape optimization of RC flexural members. *Journal of Structural Engineering*, *125*(12), 1439–1446. doi:10.1061/(ASCE)0733-9445(1999)125:12(1439)

Sadollah, A., Bahreininejad, A., Eskandar, H., & Hamdi, M. (2012). Mine blast algorithm for optimization of truss structures with discrete variables. *Computers & Structures*, *102*, 49–63. doi:10.1016/j.compstruc.2012.03.013

Sahab, M. G., Ashour, A. F., & Toropov, V. V. (2005). A hybrid genetic algorithm for reinforced concrete flat slab buildings. *Computers & Structures*, *83*(8-9), 551–559. doi:10.1016/j.compstruc.2004.10.013

Sahab, M. G., Ashour, A. F., & Toropov, V. V. (2005). Cost optimisation of reinforced concrete flat slab buildings. *Engineering Structures*, *27*(3), 313–322. doi:10.1016/j.engstruct.2004.10.002

Saini, B., Sehgal, V. K., & Gambhir, M. L. (2006). Genetically optimized artificial neural network-based optimum design of singly and doubly reinforced concrete beams. *Asian Journal of Civil Engineering (Building and Housing)*, *7*(6), 603-619.

Saka, M. P. (2009). Optimum design of steel sway frames to BS5950 using harmony search algorithm. *Journal of Constructional Steel Research*, *65*(1), 36–43. doi:10.1016/j.jcsr.2008.02.005

Schutte, J. F., & Groenwold, A. A. (2003). Sizing design of truss structures using particle swarms. *Structural and Multidisciplinary Optimization*, *25*(4), 261–269. doi:10.100700158-003-0316-5

Šešok, D., & Belevičius, R. (2008). Global optimization of trusses with a modified genetic algorithm. *Journal of Civil Engineering and Management*, *14*(3), 147–154. doi:10.3846/1392-3730.2008.14.10

Sheikholeslami, R., Khalili, B. G., Sadollah, A., & Kim, J. (2016). Optimization of reinforced concrete retaining walls via hybrid firefly algorithm with upper bound strategy. *KSCE Journal of Civil Engineering, 20*(6), 2428–2438. doi:10.100712205-015-1163-9

Sirca, G. F. Jr, & Adeli, H. (2005). Cost optimization of prestressed concrete bridges. *Journal of Structural Engineering, 131*(3), 380–388. doi:10.1061/(ASCE)0733-9445(2005)131:3(380)

Sonmez, M. (2011). Artificial Bee Colony algorithm for optimization of truss structures. *Applied Soft Computing, 11*(2), 2406–2418. doi:10.1016/j.asoc.2010.09.003

Talatahari, S., & Kaveh, A. (2015). Improved bat algorithm for optimum design of large-scale truss structures. Iran University of Science & Technology, 5(2), 241-254.

Talatahari, S., Sheikholeslami, R., Shadfaran, M., & Pourbaba, M. (2012). Optimum design of gravity retaining walls using charged system search algorithm. *Mathematical Problems in Engineering*, 2012.

Temür, R., & Bekdaş, G. (2016). Teaching learning-based optimization for design of cantilever retaining walls. *Structural Engineering and Mechanics, 57*(4), 763–783. doi:10.12989em.2016.57.4.763

Temür, R., Kayabekir, A. E., Bekdaş, G., & Nigdeli, S. M. (2018). Grey Wolf Optimizer Based Design of Reinforced Concrete Retaining Walls Considering Shear Key. In *7th International Conference on Applied and Computational Mathematics (ICACM '18)*. Rome, Italy. Academic Press.

Toğan, V. (2012). Design of planar steel frames using teaching–learning based optimization. *Engineering Structures, 34*, 225–232. doi:10.1016/j.engstruct.2011.08.035

Toğan, V., & Daloğlu, A. T. (2008). An improved genetic algorithm with initial population strategy and self-adaptive member grouping. *Computers & Structures, 86*(11-12), 1204–1218. doi:10.1016/j.compstruc.2007.11.006

Ulusoy, S., Kayabekir, A. E., Bekdaş, G., & Nigdeli, S. M. (2018) Optimum Design of Reinforced Concrete Multi-Story Multi-Span Frame Structures under Static Loads. In *8th International Conference on Environment Science and Engineering (ICESE 2018)*, Barcelona, Spain. Academic Press.

Venkayya, V. B. (1971). Design of optimum structures. *Computers & Structures*, *1*(1-2), 265–309. doi:10.1016/0045-7949(71)90013-7

Yepes, V., Alcala, J., Perea, C., & González-Vidosa, F. (2008). A parametric study of optimum earth-retaining walls by simulated annealing. *Engineering Structures*, *30*(3), 821–830. doi:10.1016/j.engstruct.2007.05.023

Zielinskil, Z. A., Long, W., & Troitsky, M. S. (1995). Designing reinforced concrete short-tied columns using the optimization technique. *Structural Journal*, *92*(5), 619–626.

Chapter 2
Brief Information About Metaheuristic Methods:
History and Theory of Metaheuristics

ABSTRACT

First, the reasons of using metaheuristic algorithms are listed for optimum design of reinforced concrete (RC) structures and members. The main reason is the non-linear formulation of RC design problems including various types of design constraints. The basic terms and formulation of optimization methods are presented. A general process of metaheuristic-based optimization methods is presented. This process is summarized as three stages: pre-optimization, analysis, and optimization. Details of several metaheuristic algorithms effectively used in structural engineering problems are summarized by giving all formulations according to the inspiration of algorithms. The flowcharts for the optimization processes are also included.

THE REASONS OF USAGE OF METAHEURISTIC METHODS

Conventional methods can be used to solve optimization problems, but engineering problems generally have high numbers of design constraints. In that case, the solution of optimum design variables cannot be directly calculated since these variables are effective on the analyses of design constraints. This situation makes most of the engineering problem non-linear.

DOI: 10.4018/978-1-7998-2664-4.ch002

The easiest way to solve a non-linear problem is to use trials with iterative solutions, but this process may last too long. For that reason, the randomization and choosing of candidate design variables must be done in a systematic way by using methodologies employing iterative algorithms. These iterative algorithms are mainly metaheuristic methods. The development of metaheuristic methods includes a process which has the same goal for objective of an optimum design. Every process has objectives to provide or objective functions to minimize or maximize.

Especially, the optimum design of reinforced concrete (RC) structures is more complex than other engineering problems because of following reasons:

1. RC member consists of two different materials such as concrete and steel. These materials have totally different strength behavior and cost.

Steel has the same strength under tensile and compression. Concrete is a brittle material and has an acceptable compressive strength while the tensile strength is neglected. For the protection of steel from fire and environmental conditions, concrete is an excellent cover with low cost.

If a part of a section of RC member is under tensile, reinforcing bars (rebar) must be added. For civil structures, ductile design must be developed. In other words, when the yielding strength is exceeded, the structure must continue to carry the internal forces resulting from static and dynamic forces. Steel is a ductile material, but this situation cannot be provided with brittle material, concrete. For that reason, several rules are given in design codes. These rules are separately mathematically modelled as design constraints in optimization.

The most basic ductile behavior rule for members under flexural moments is to provide yielding of steel before the fracture of concrete. In that case, the amount of steel is limited. To carry more forces, the parts under compressive stress must also be reinforced with rebar. In that case, candidate solutions may include both reinforcements in tensile and compressive sections, or only reinforcements can be positioned in tensile section. For that reason, the code of optimization methodology must be considering this and similar situations.

2. RC structures contain several types of members which are under different types of stresses. For example, beam is dominantly under flexure effects while columns are dominantly under an axial force. Foundations and

retaining walls are under the effect of geotechnical constraints in addition to structural constraints.

3. Different members of same types are under effects of forces with different intensities. In that case, each member has a separate optimum design. Because of statically undetermined systems, the change of rigidity effects the solution.
4. The concrete sections are casted in construction yard. Precise dimensions cannot be provided. This situation leads us to use discrete design variables by uncompromising a precise optimum design.
5. In local market, fixed sizes of rebar are found. In that case, the available sizes must be used in random optimization.
6. Structures are huge systems and load on structures are not certain. The self-weight of structural and non-structural members are known, but the amount of live load and existing of live load in a span is not known and it is changeable. Additionally, dynamic loads are totally unknown. The earthquake accelerations and wind speed cannot be predicted. Only vibration periods and maximum intensities can be guessed. Additionally, a region may have special forces (earthquake, wind, snow, etc.)
7. In addition to specific loads, the requirements of a country may be different because of the usage of different design codes. Also, the constructed structure may have different soil condition and usage purposes.

By coding the best suitable metaheuristic and development of a specific methodology, these difficulties about optimization of RC members are solved.

BRIEF HISTORY OF METAHEURISTICS

In recent years, a lot of metaheuristic algorithms have been developed and it is not possible to count the number of algorithms. Although, the developments of metaheuristic algorithms were done in recent years by using a name coming from inspiration (metaphor), the idea of metaheuristic dates back before 1940s according to Sörensen, Sevaux and Glover (2018). The brief information about following periods can be found in this chapter.

1. The pre- theoretical period (until 1940).
2. The early period (1940-1980).
3. The method-centric period (1980-2000).
4. The framework-centric period (2000-now).

5. The scientific period (future).

The Pre-Theoretical Period (Until 1940)

Optimization exist in all problems of life from the start of human life. Even in easy life tasks like finding the shortest path, optimization is used by the human mind. By the born of a baby, human mind is equipped to solve a great variety of problems and the human mind is a system which is capable to use metaheuristic strategies. As a basic example, human searches past solutions of similar problem to find new rules for their problems. According to Sörensen et al. (2018), heuristic and even the improved meta form of heuristic (metaheuristics) are natural to human.

The Early Period (1940-1980)

After the World War II, printed documents about strategies which can be used in development of optimization algorithms had been published. The book developed by Polya called "How to Solve it" (Polya, 2004) is one of the latest examples. The following principles used in the development of heuristic were mentioned in this period.

1. Analogy: To investigate another solved problem close to current interest.
2. Induction: To use generalization of some examples.
3. Auxiliary problem: To use a sub-problem.

These principles are handled together in metaheuristic. In the development of metaheuristics, the steps are as follows:

1. Find the other problem (analogy).
2. Formulize the learned techniques of problem (induction).
3. Decompose to smaller problems by separating different techniques as different optimization stages (auxiliary problem).

In this period, the greedy rules used for selection of best value were also suggested (Cormen, Leiserson, Rivest, & Stein, 2009). Also, "ill-structured" problems which cannot be formulated or solved by classical method easily are mentioned by Simon and Nevell (1958).

The Method-Centric Period (1980-2000)

The early metaheuristics were mentioned in this period. Although evolutionary algorithms had been developed around 60s, the true start is at this period. The main idea in this period is to find the optimum solution of a problem. For this reason, several metaheuristic algorithms were developed by using metaphors. Some of the early metaheuristics which are used in engineering problems are as follows:

1. Tabu Search (TS) using the human memory (Golver, 1986).
2. Simulated Annealing (SA) using process of annealing in material production (Kirkpatrick, Gelatt, & Vecchi, 1983).
3. Particle Swarm Optimization (PSO) using the movement of swarms including birds, fishes, etc. (Kennedy & Eberhart, 1995).
4. Ant Colony Optimization (ACO) using the behaviour of ants seeking a path (Dorigo, Maniezzo, & Colorni, 1996).
5. Differential evolution using mutation, crossover and selection (Storn & Price, 1997).

The algorithm developed in this period contains the major features of metaheuristic algorithm and the ones developed after these algorithms contain similar features that are modified.

The Framework-Centric Period (2000-now)

During this period, the framework became more important than the method. Several new methods, modifications of these methods and hybrid methods have been developed, and the purposes of these methods are as follows:

1. Reach optimum solution quicker than the others.
2. Robustness of results for several runs of algorithms.
3. Minor improvements in objective function.
4. Easy application of the developed method.

According to Sörenson et al. (2018), a sub-period called as metaphor-centric period exists and these methods are criticized (Sörensen, 2015, Weyland, 2010) since these methods are a copy of the existing features.

Rajpurohit, Sharma, Abraham and Vaishali (2017) developed a glossary of metaheuristic methods. In this glossary, the metaphors of metaheuristic methods are listed. In the Table 1, several metaheuristic algorithms, which are often used in structural engineering problems, are given with their metaphor.

The Scientific Period (Future)

Metaheuristics will remain important for the future since the nature of non-linear optimization problems are not suitable for other types of methods. To provide more income from the use of metaheuristics, researchers must

Table 1. The metaheuristic algorithms developed in framework-centric period

Algorithm	Developer	Year	Metaphor
Harmony Search	Geem, Kim and Loganathan	2001	Process of finding the best harmony to gain the appreciation of the audience
Big bang-big Crunch	Erol and Eksin	2006	Big-bang big-crunch theory used for evolution of the universe
Artificial Bee Colony Algorithm (ABC)	Karaboga and Basturk	2007	Intelligent foraging behavior of honey bee swarm
Cuckoo Search Algorithm	Yang and Deb	2010	Inspired from brood parasitic behavior of some cuckoo species
Firefly Algorithm	Xin-She Yang	2010	Imitates flashing light property of fireflies
Bat Algorithm	Xin-She Yang	2010	Echolocation behavior of bats used for search directions and location of insects
Teaching-learning based Optimization	Rao, Savsani and Vakharia	2011	Simulation of teacher-learner relationship during education process
Flower Pollination Algorithm	Xin-She Yang	2012	Pollination process of flowering plants
Ray Optimization	Kaveh and Khayatazad	2012	Refraction property of light rays
Krill Herd	Gandomi and Alavi	2012	Process of the herding behavior of krill
Grey Wolf Optimizer	S. Mirjalili, S.M. Mirjalili and Lewis	2014	Hierarchical leadership and hunting methods of grey wolves
Jaya Algorithm	R. V. Rao	2016	Sanskrit word mean victory

provide more information about the coding of the algorithm and development methods must be quicker than the existing ones. Thus, the problems of science can be clearly solved.

METHODOLOGY OF METAHEURISTIC METHODS

Formulation of Optimization Problem

In optimization problems, there are several indexes as follows:

1. Design constants.
2. Design variables.
3. Objective functions.
4. Design constraints.

The main aim of optimization problems is to provide best suitable values of "design variables" providing minimization or maximization of single or multiple "objective function" for a problem defined with fixed and known values called as "design constants" by providing several inequalities ($g_k(x)$) and equality ($h_j(x)$) functions defined as "design constraints". A set of design variables (x) can be shown as Eq. (1) for a problem with n design variables.

$$x=(x_1, x_2, \ldots, x_i, \ldots x_n)^T \text{ for } i=1, 2, \ldots, n \tag{1}$$

The design variables are variables of 'a' numbers of objective functions ($f_m(x)$), 'b' numbers of equality functions ($h_j(x)$) and 'c' numbers of inequality function ($g_k(x)$) as shown in Eqs. (2)-(4).

$$f_m(x), x \in \mathbb{R}^n, (m=1, 2, \ldots, a) \tag{2}$$

$$h_j(x)=0, (j=1, 2, \ldots, b) \tag{3}$$

$$g_k(x) \leq 0, (k=1, 2, \ldots, c) \tag{4}$$

In structural engineering, design variables are generally dimension of the system, amount of used material (for example diameter of rebar) and material or system properties. In an optimization problem, different types of variables can be considered, and these variables may be continuous or discrete, while a single problem may contain mixed variables.

The design constants are generally design limits provided by constructor, allowable material properties, soil conditions, environmental conditions, etc. Some values can be taken as constant, while these values may be design variables for another problem or case of optimization.

The design constraints in structural engineering problem are the rules defined in design codes according to theory of mechanics or architectural constraints can be also provided according to designer demands.

The objective functions of structural engineering problems may be minimization of cost, reduction of internal forces, minimization of CO_2 emission in production, increasing the performance of an additional system (i.e. tuned mass dampers, viscous dampers in dynamic structures). In several cases, several factors can be handled as an objective or a design constraint.

Fundamental Methodology of Metaheuristic Methods

In this section, a general methodology of metaheuristic methods is given. The detailed methodology of several methods is explained in the following sections of this chapter.

The main idea of metaheuristics is generating candidate solutions, modifying these solutions and selection of the best one for the optimization objectives. For a coded methodology shown in Fig. 1 as a flowchart, the process starts with definition of design constants. Together with the design constants, the range of design variables and the algorithm-specific parameters must be defined. Population based algorithms have several sets of design variables and the population (p) must be defined.

Then, an initial solution matrix containing set of design variables is generated. The initial values are generally randomly chosen from the solution range with minimum (x_{imin}) and maximum (x_{imax}) limits as seen in Eq. (5) for generation of ith design variable of jth individual of population. rand (1) is random number defined between 0 and 1. In optimization problems, there are several indexes as follows:

$$x_i^j = x_{imin} + rand(1)(x_{imax} - x_{imin}) \text{ for } i=1, 2, \ldots, n \text{ and } j=1, 2, \ldots, p \quad (5)$$

After pre-optimization stages, the objective functions are calculated and saved for future comparison of modified values. The design constraints

Figure 1. A general flowchart of metaheuristic algorithm

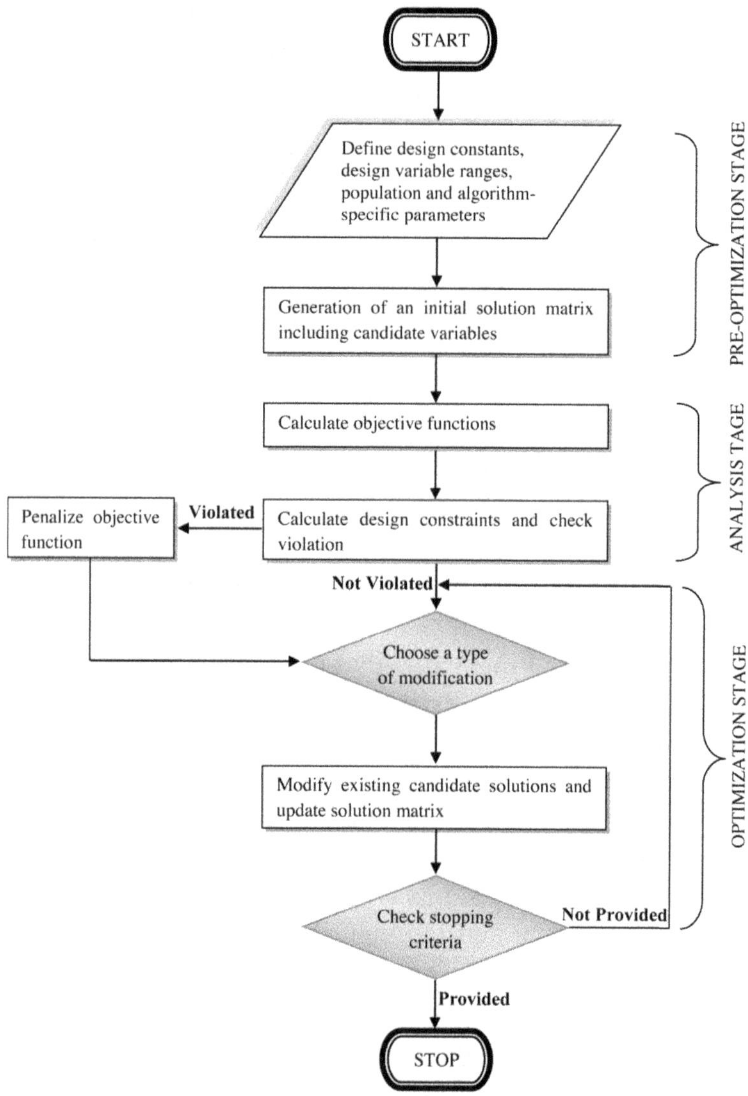

are also checked. If single or multiple design constraints are violated, the objective functions are penalized by adding an increasing value for minimizing objectives. For maximized solution, penalty function is taken as a reduction of objective function. These steps are mentioned as analysis stage of the flowchart.

Then, the essential optimization starts, and the special formulation of metaheuristic algorithm formulated according to inspiration features are applied. In the formulation of modification of design variables, several algorithm specific parameters may be used. Also, several phases of optimization can exist. These phases are generally chosen according to an algorithm parameter. After the modification of existing design variables, the objective functions are checked by considering design constraints. If an improvement in solution of objective function is seen, the existing results are changed with newly modified ones. This process iteratively continues until stopping criteria are provided.

The mostly used stopping criterion is to evaluate a constant number of iterations. In that case, a maximum iteration number must be defined. Differently from this criterion, an amount of reduction of objective function, convergence of objective function, etc. can be also chosen.

The optimization stages of algorithms are briefly explained in the following sections with formulations and specific flowcharts.

BRIEF INFORMATION ABOUT SEVERAL METAHEURISTIC ALGORITHMS

In this section, several metaheuristic algorithms mentioned in the past studies about optimum design of RC structures and members are briefly summarized. A flowchart for each algorithm is presented for optimization stage of general flowchart shown as Fig.1.

Genetic Algorithm

The main idea used in metaheuristic algorithms is the evolution theory. The features of biological evolution are reproduction, mutation, recombination and selection. In an iterative optimization, similar features are generally used. The examples are as follows:

- Reproduction: In the iterations, we need to produce new candidate design variables.
- Mutation: In generation of candidate solutions, existing solutions are needed to be used to reach better solutions with a good convergence.
- Recombination: Newly generated solutions may be better or worse than the existing ones. In that case, better solutions according to objective functions must be stored.

- Selection: The optimization processes involve several types of selection and decision making. For example, the type of generation of design variables can be chosen. Or in a modification formulation, random selection of existing solutions and best or worst solutions are needed.

Most of the developed metaheuristic algorithms are also mentioned as evolutionary algorithms since a biological metaphor or features of evolution exist in these algorithms, but genetic algorithm (GA) is the earliest algorithm imitating evolutionary features. Mainly, GA uses crossover, mutation and selection in the evolution theory of Charles Darwin and it was firstly mentioned by John Holland (Holland, 1975). Then, the first structural optimization attempt with GA was done by Goldberg and Samtani (1986). In the first mentioned classical GA, encoding and decoding is in need. For example, a task will be accepted or rejected. The accepted ones are shown as 1, while 0 is assigned for the rejected tasks. For these tasks taken as design variables, a population of chromosome including genes regarded as tasks are encoded as shown in Fig. 2.

In Fig. 2, a problem using 3 numbers of chromosomes as population and 5 genes as design variables is represented. For these chromosomes including values of candidate design variables, the objective function or namely, fitness function is calculated. For example, each gene has a value if accepted. By order, selection, crossover and mutation processes will be applied in an iteration.

In selection, two parent chromosomes are chosen. A method or random selection can be used to find parents. Then, two chosen parents are used in crossover to generate off-springs. For example, first and second chromosomes are chosen as parents and the off-springs will be as seen in Fig. 3.

Then, the mutation stage starts, because a child may not be the exact copy of parents.

In the final, the newly generated child chromosomes are replaced with the chromosomes with the lowest fitness function.

Figure 2. The population in GA

					Gene	
Chromosome 1	1	0	0	1	0	Population
Chromosome 2	0	1	1	0	1	
Chromosome 3	1	1	1	1	0	

Figure 3. Crossover in GA

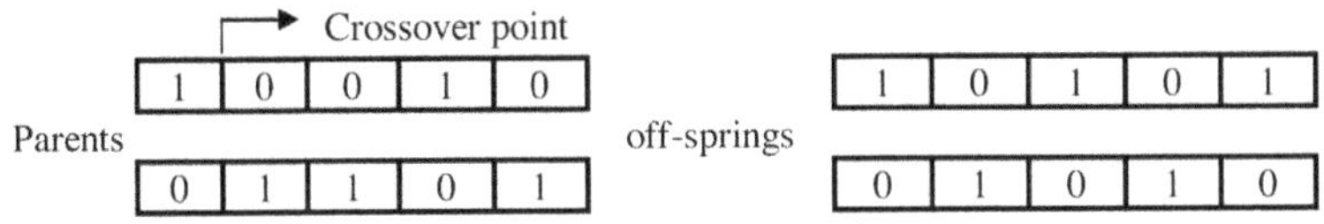

Figure 4. Mutation in GA

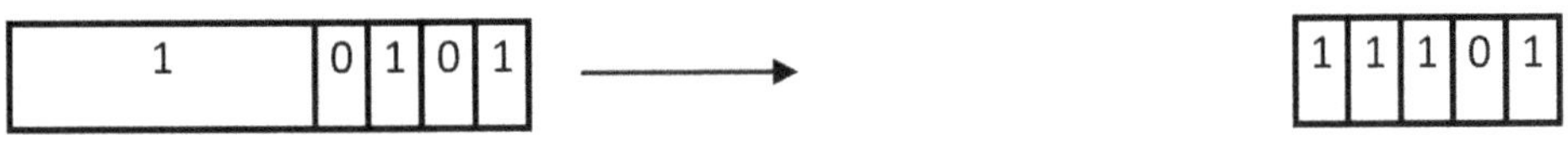

A general flowchart of GA was presented for essential optimization process as given in Fig. 5. For current design optimization problems, GA is used with modification considering the fundamental features.

Differential Evolution Algorithm

The evolution features such as mutation, crossover and selection were used in the development of differential evolution (DE) algorithm by Storn and Price (1997). DE was developed to outcome the following user demands:

- Ability to solve non-differentiable, nonlinear and multimodal objective functions.
- Coping with analyses of the intensive objective functions with design stages.
- Being a user-friendly algorithm with less and robust parameters.
- Providing a good convergence to save computational time.

The formulation of DE includes generation of mutated vector. For i^{th} value of n design variables, j^{th} individual of population (p) and $t+1^{th}$ iteration of t_{max} maximum iteration, the mutated value ($v_i^{j,t+1}$) is calculated as Eq. (6).

$$v_i^{j,t+1} = x_i^{a,t} + F(x_i^{b,t} - x_i^{c,t}) \text{ for } i=1, 2, \ldots, n \text{ and } j=1, 2, \ldots, p \text{ and } t=1, 2, \ldots, t_{max} \quad (6)$$

In Eq. (6), the values; a, b and c are randomly defined individual numbers (between 1 and p) and $x_i^{a,t}$, $x_i^{b,t}$ and $x_i^{c,t}$ define the randomly selected values

Figure 5. Flowchart of GA

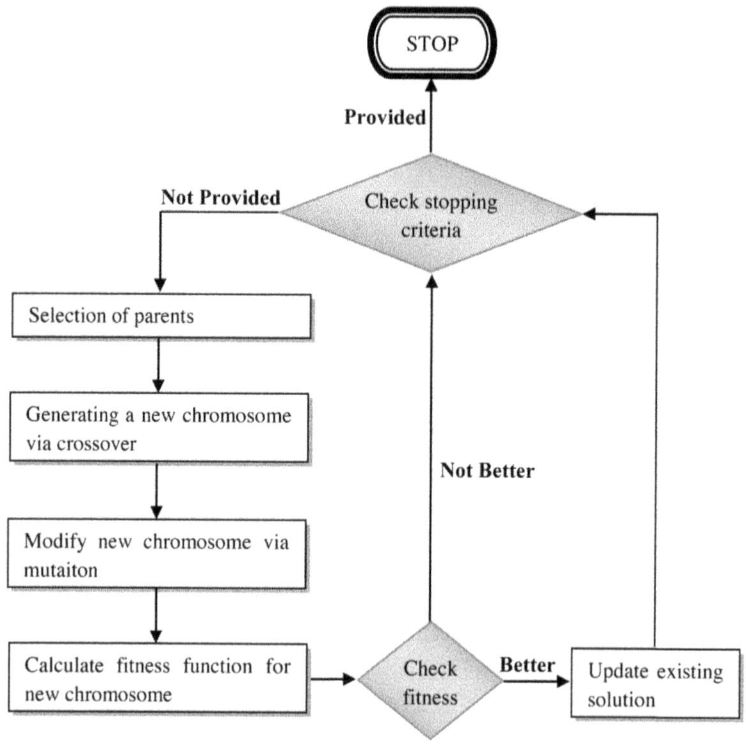

of tth iteration and i[th] design variable. F is a control parameter to adjust the amplitude of $x_i^{b,t} - x_i^{c,t}$. It is taken as a real constant number between 0 and 2.

After the generation of mutated vectors ($v^{1,t+1}$, $v^{2,t+1}$, …, $v^{p,t+1}$), a crossover process with the existing vector ($x^{1,t}$, $x^{2,t}$, …, $x^{p,t}$) is done to generate trial solution vectors, which are the final vectors of next iteration ($x^{1,t+1}$, $x^{2,t+1}$, …, $x^{p,t+1}$). This process is formulated as Eq. (7).

$$x^{j,t+1} = \begin{cases} v^{j,t+1}\ if\ rand(1) \leq CR\ or\ j = r \\ x^{j,t}\ if\ rand(1) > CR\ or\ j \neq r \end{cases} \tag{7}$$

A random number assigned between 0 and 1 is defined as rand(1). The crossover constant is shown as CR and it is also assigned between 0 and 1 by user. The value; r is an integer number randomly chosen between 1 and p. The reason of using a random index is to ensure that at least one of the mutated vectors is chosen. Otherwise, all solutions may be the same as the existing one.

In the final stage named as selection, the existing results ($x^{j,t}$) and trial results ($x^{j,t+1}$) are compared according to the objective function and the best one is stored as the final solution of $(t+1)^{th}$ iteration.

The stages such as mutation, crossover and selection are shown in the flowchart presented as Fig. 6.

Particle Swarm Optimization

The behaviour of swarms is one of the most common inspirations of metaheuristic algorithm. It firstly formulized as particle swarm optimization (PSO) and behaviours of specific natural happening are coded as optimization algorithms after development of PSO. The detailed information about swarm intelligence can be found in Parpinelli and Lopez (2011).

Differently from evolutionary features such as mutation and crossover, swarm-based algorithms use simplified features by direct formulation of natural happenings.

In PSO developed by Kennedy and Eberhart (1995), a position vector is created and it is added to existing solution as seen in Eq. (8).

$$x_i^{j,t+1} = x_i^{j,t} + v_i^{j,t+1} \text{ for } i=1, 2, \ldots, n \text{ and } j=1, 2, \ldots, p \text{ and } t=1, 2, \ldots, t_{max} \quad (8)$$

Figure 6. Flowchart of DE

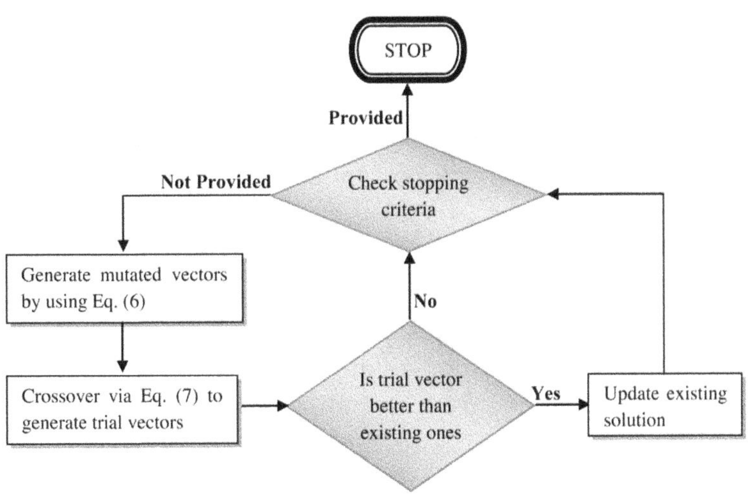

$v_i^{j,t+1}$ represents position value of ith design variable, $(t+1)^{th}$ iteration and j^{th} individual and it is formulated as Eq. (9).

$$v_j^{j,t+1}=\theta(t)v_i^{j,t} + \alpha(rand(1))(g_i^*-x_i^{j,t})+\beta(rand(1))\,(x_i^{*,t}-x_i^{j,t}) \quad (9)$$

In Eq. (9), two learning parameters such as α and β are used. These parameters are suggested to take as 2 in value. An inertia function; $\theta(t)$ respect to iteration (t) is proposed to use to control the weight of the velocity. It may be a function or a constant value. The global best optimum solution of ith design variable is shown as g_i^*. This value is the best of all solutions within past iterations. $x_i^{*,t}$ is the local best solution and it is the best solution of the last generated values (best of $x_i^{1,t}$, $x_i^{2,t}$, ..., $x_i^{p,t}$).

After each iteration, the existing ones and newly generated solution are compared according to the objective function and the best ones are only accepted. The flowchart of PSO is shown as Fig. 7.

Flower Pollination Algorithm

Yang (2012) developed a metaheuristic algorithm by imitating the pollination process of flowering plants. In the development of the algorithm, specific types of flower pollination and flower constancy are the key features in generation of a metaheuristic algorithm.

As known, pollination is the process of reproduction of flowering plants. Flower constancy is the feature which involves a specialised flower-pollinator partnership. In this partnership, a specific flower only attracts specific pollinators. The constancy is the key feature in the development of the algorithm and this idea is combined with the pollination types to provide global and local pollination stages. Global and local pollination stages are chosen by controlling a switch probability (sp). For example, sp is chosen as 0.5. A random number between 0 and 1 is generated. If generated number is lower than sp, global optimization will be applied. Otherwise, local optimization will be in progress.

Global pollination covers two types of pollinations, because pollination has types for different of pollinators and flower. According to difference of pollinator, biotic pollination is the inspiration of global pollination or global optimization. In biotic pollination, living bio-organism are responsible with pollinations. These bio-organisms as pollinators are insects, bees or other animals and 90% of flower pollination is done via biotic pollination. Pollinators can transfer pollen for long distances by flying. In that case, biotic pollination is used as global pollination. The other type related with flower species is cross pollination. Cross pollination is the type in which the pollen transfer can be done between different flowers. In that case, the candidate flowers form the high amount of all flowers.

In global pollination, an updated solution of next iteration ($x_i^{j,t+1}$ for i^{th} design variable, j^{th} flower and $(t+1)^{th}$ iteration) can be calculated as follows:

$$x_i^{j,t+1} = x_i^{j,t} + L(x_i^{j,t} - g_i^*) \text{ for } i=1, 2, \ldots, n \text{ and } j=1, 2, \ldots, p \text{ and } t=1, 2, \ldots, t_{max} \quad (10)$$

In Eq. (10), L denotes a Lévy distribution which represents a random flight. Pollinator obey the rules of Lévy flight. g_i^* denotes the best existing result within the generated matrix.

Local optimization imitates abiotic and self-pollinations. In abiotic pollination, the pollen transfer process is done via wind or diffusion in water without use of living bio-organisms. In self-pollination, self-fertilization of flower can be also done without a pollinator like peach. Self-pollination

Figure 7. Flowchart of PSO

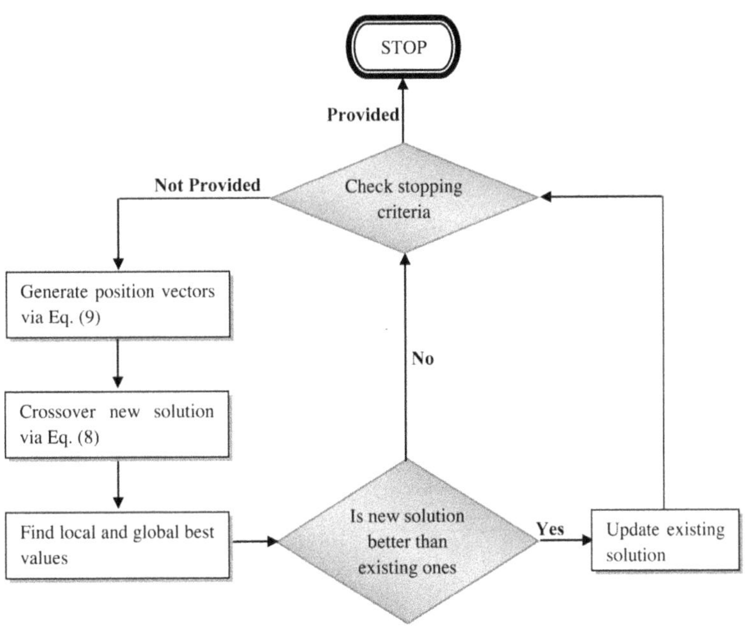

covers only one type of plant. Because of these reasons, abiotic and self-pollinations are mentioned as local pollination. In FPA, local optimization is formulated as Eq. (11).

$$x_i^{j,t+1} = x_i^{j,t} + \varepsilon(x_i^{a,t} - x_i^{b,t}) \text{ for } i=1, 2, \ldots, n \text{ and } j=1, 2, \ldots, p \text{ and } t=1, 2, \ldots, t_{max} \quad (11)$$

In Eq. (11), a linear distribution (ε) is used instead of a Lévy distribution. This linear distribution is generally a random number between 0 and 1. In local optimization, two randomly selected existing values are used ($x_i^{a,t}$ and $x_i^{b,t}$) . a and b denote selected individuals in the pollination.

Newly generated solutions are updated if new ones are better than existing ones for the solution of objective function. The flowchart of FPA for specific part of the algorithm is given as Fig. 8.

Bat Algorithm

In Bat Algorithm (BA) developed by Yang (2010), echolocation behaviour of micro bats is idealized as a metaheuristic algorithm. In the echolocation

Figure 8. Flowchart of FPA

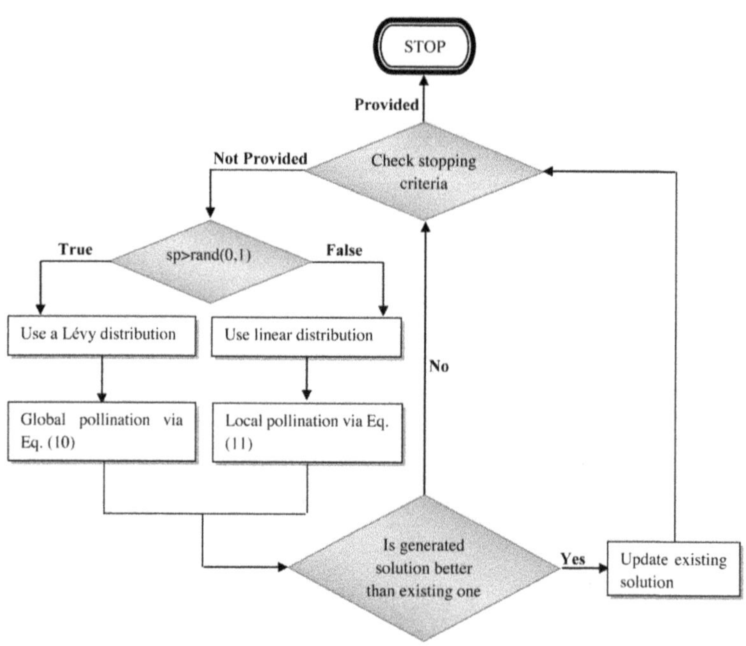

behaviour of micro bats, bats tune a specific frequency, when they home to their prey. During tuning of frequency, varying of several factors such as frequency, loudness and pulse emission rates is in progress. In this situation, an optimization is done by bats. For that reason, the randomization idea can be used in the development of a metaheuristic algorithm.

The location of bats is defined with a location vector and a location; $x_i^{j,t+1}$ for ith design variable, jth bat and (t+1)th iteration is formulated as Eq. (12).

$$x_i^{j,t+1}= x_i^{j,t} +v_i^{j,t+1} \text{ for } i=1, 2, \ldots,n \text{ and } j=1, 2, \ldots,p \text{ and } t=1, 2, \ldots,t_{max} \tag{12}$$

$v_i^{j,t+1}$ is the velocity of jth bat for ith design variable and (t+1)th iteration. It is formulated as Eq. (13).

$$v_i^{j,t+1}= v_i^{j,t} + (x_i^{j,t}-g_i^*)f_i \tag{13}$$

In equation of velocity, the best existing solution of ith design variable (g_i^*) and a varying frequency of bats (f_i) are used. The frequency is randomized between minimum (f_{min}) and maximum (f_{max}) frequency as shown in Eq. (14).

$$f_i= f_{min} + (f_{max}-f_{min})(rand(1)) \tag{14}$$

After the generation of location ($x_i^{j,t+1}$), candidate solutions of these vectors are accepted or not instead of existing solution. For this decision, a criterion test is applied according pulse rate (r_i^t) and loudness (A_i^t).

If pulse rate is smaller than a generated random number, local search is used to reproduce location vector. In local search, loudness is used as formulated in Eq. (15).

$$x_i^{j,t+1}= x_i^{j,t} + \varepsilon A_i^t \tag{15}$$

In local search, a random walk function (ε) is used and ε is a random number between -1 and 1.

The pulse rate and loudness are varying according to iterations. These algorithm variables vary as formulated in Eqs. (16) and (17).

$$A_i^{t+1}= \alpha A_i^t \tag{16}$$

$$r_i^{t+1} = r_i^0[1\text{-}\exp(-\gamma^t)] \quad (17)$$

In the Eqs. (16) and (17), α and γ^t are algorithm constants and the values are used to control range of local search. r_i^0 is the user defined initial pulse rate value. Like all algorithms, better results than existing ones are selected and saved.

The flowchart of the essential optimization process of BA is given as Fig. 9.

Simulated Annealing

Simulated Annealing (SA) inspired from the annealing process in metallurgy. In the annealing process, heating and controlled cooling of a material are done to increase the size of crystals of material. Thus, defects of materials are reduced.

The algorithm; SA proposed by Kirkpatrick et al. (1983) is a search method using Markov chain to prevent local optima trapping. Markov chain has the ability of convergence under appropriate conditions by considering not only improved results. Some results which are not better than existing ones are also accepted.

Firstly in SA, initial temperature (T_0) and initial guess of design variables ($x_1^0, x_2^0, \ldots, x_n^0$ for n number of design variables) are done. During iterations, the generation of candidate design variables are done as Eq. (18) by using a pseudorandom number (randn).

$$x_i^{t+1} = x_i^t + \text{randn for } i=1, 2, \ldots, n \text{ and } t=1, 2, \ldots, t_{max} \quad (18)$$

SA is not a population-based method. Only single values of design variables are assigned and saved after iterations.

The generation of candidate solutions is done until the temperature (T) is lower than the final temperature and iteration number (t) is lower than maximum number of iterations (t_{max}). Temperature can be defined with linear (Eq. (19)) or geometric (Eq. (20)) functions.

$$T(t) = T_0 \text{-}\beta t \quad (19)$$

$$T(t) = T_0 \alpha t \quad (20)$$

Figure 9. Flowchart of BA

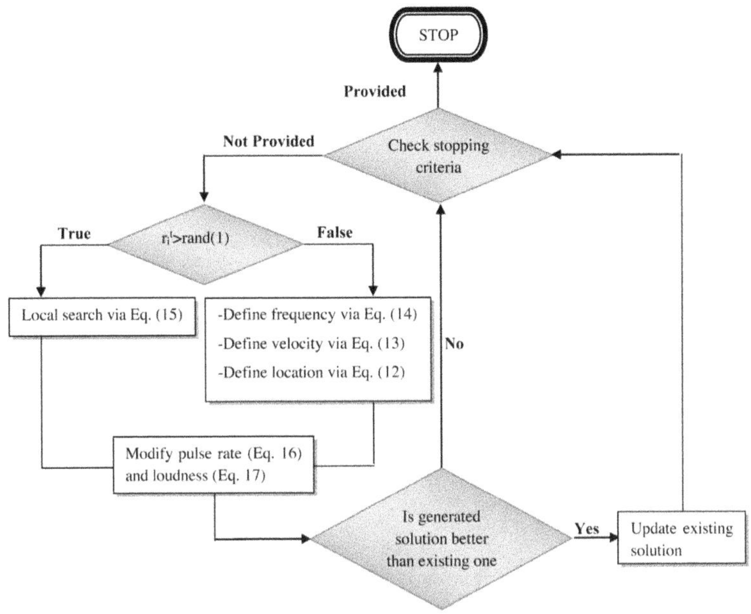

The cooling rate (β) and cooling factors (α) are used in calculation of temperature. The generated candidate results are accepted according to following two criteria. The results providing one of these factors are accepted.

- The value of objective function for new candidate values (x^{t+1}) must be better than the value of existing ones (x^t).
- The inequality presented as Eq. (21) must be provided. Δf is the difference of the objective function values of the last iterations ($f(x^{t+1})$ and $f(x^t)$) as formulated in Eq. (22). r is a randomly defined number.

$$\exp[-\Delta f/T]>r \tag{21}$$

$$\Delta f = f(x^{t+1}) - f(x^t) \tag{22}$$

The flowchart of SA is given as Figure 10.

Figure 10. Flowchart of SA

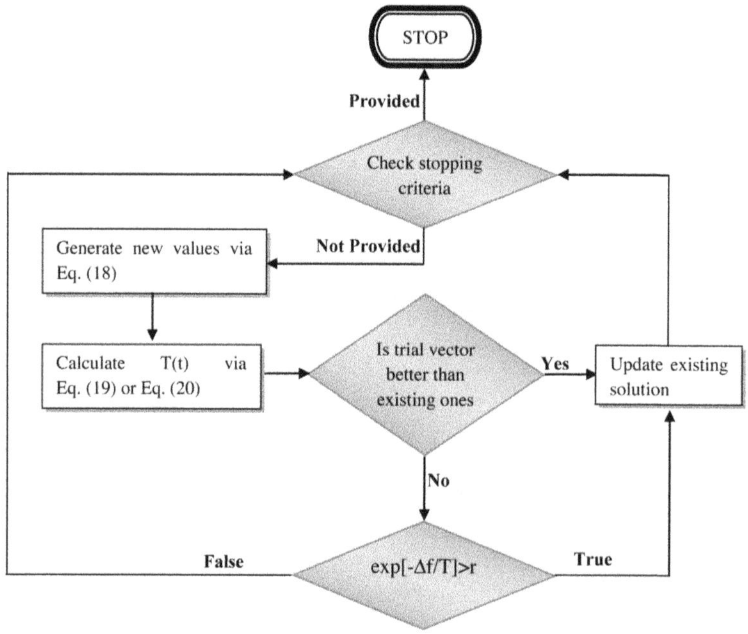

Teaching-Learning-Based Optimization

Rao et al. (2011) developed teaching-learning-based optimization (TLBO) algorithm and two phases of education are formulated. These phases are the teacher phase in which the education is carried out by a member with the best knowledge and the learner phase in which self-study of the classroom is done via knowledge transfer between students.

The formulation of teacher phase in development of a candidate solution ($x_i^{j,t+1}$) is given as Eq. (23)

$$x_i^{j,t+1} = x_i^{j,t} + rand(0,1)(g_i^* - TFx_i^{ave}) \text{ for } i=1, 2, \ldots, n \text{ and } j=1, 2, \ldots, p \text{ and } t=1, 2, \ldots, t_{max} \quad (23)$$

Since the education is done by a teacher and teacher is the best person with the knowledge, the best existing solution (g_i^*) is used in the teacher phase. The aim is to increase capacity of the average of the class and the average of the existing solution (x_i^{ave}) is used. A teaching factor (TF) is used to control the weight of the average, but TF is not a user defined parameter. It is randomly chosen as 1 or 2. In that case, TLBO is a user-friendly algorithm since user-defined parameters are not needed.

The teacher phase is considered as global optimization. Differently from the most of metaheuristic algorithms, TLBO has no probability parameter to choose between global and local search. Both phases or types are consequently done in an iteration.

The learner or student phase is the local optimization part of TLBO. The formulation of learner phase is shown as Eq. (24).

$$x_i^{j,t+1} = \begin{cases} x_i^{j,t} + rand(1)(x_i^{a,t} - x_i^{b,t})\, if\, f(x_i^{a,t}) < f(x_i^{b,t}) \\ x_i^{j,t} + rand(1)(x_i^{b,t} - x_i^{a,t})\, if\, f(x_i^{a,t}) > f(x_i^{b,t}) \end{cases} \tag{24}$$

In the learner phase, like several optimization phases, two existing solutions ($x_i^{a,t}$ and $x_i^{b,t}$) of two random individuals (a and b) are used. This situation represents the self-education after teaching. All students have part of good knowledge can be expressed to others. After each consequent phase, new results are replaced with the existing ones if new ones have a better objective function. The flowchart of TLBO for essential optimization is shown as Fig. 11.

Figure 11. Flowchart of TLBO

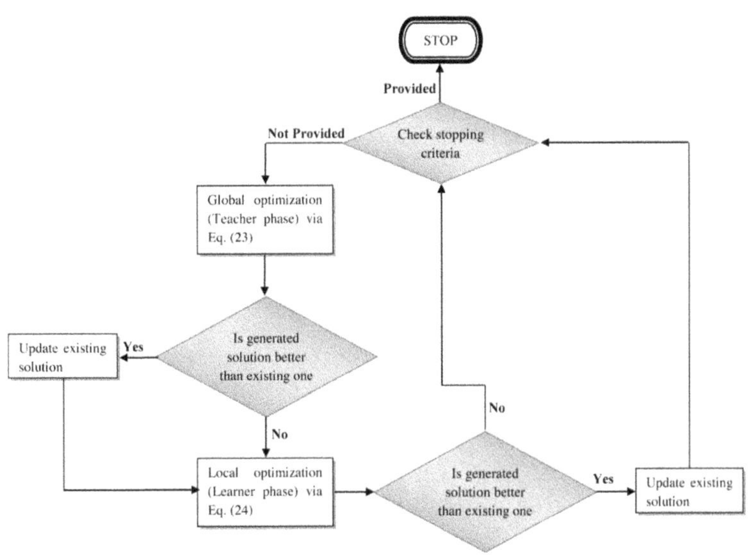

Harmony Search Algorithm

Harmony Search (HS) algorithm is a music inspired metaheuristic algorithm and it is developed by Geem, Kim and Loganathan (2001) via observation of musical performances. In musical performances, musician tries to adjust tunes according to audience to get their attention. In this process, the effort of musicians is as follows:

- Playing popular notes.
- Playing new notes.
- Playing notes that are similar to known or played ones.

By these efforts, notes are always updated to get the admiration of audience. This process has the same goal of an optimization problem. HS has two phases and two user-defined parameters. These parameters are harmony memory considering rate (HMCR) and pitch adjusting rate (PAR). HMCR is generally used as to choose from global and local optimization. PAR is used to adjust the neighbouring values. By this idea, the formulation of HS can be differently formed. In classical HS, a neighbourhood index (n_i) and an arbitrary distance bandwidth (b_w) is used in local search for discrete and continuous variables, respectively.

The steps of classical HS are as formulated in Eqs. (25-28).

$$x_i^{j,t+1} = x_{i,min} + rand(1)(x_{i,max} - x_{i,min}) \text{ if } HMCR > r_1 \quad (25)$$

$$x_i^{j,t+1} = x_i^{j,t} \text{ if } HMCR \leq r_1 \text{ and } PAR > r_2 \quad (26)$$

$$x_i^{j,t+1} = x_i^{j+n_i,t+1} \text{ if } HMCR \leq r_1 \text{ and } PAR \leq r_2 \text{ (for discrete variables)} \quad (27)$$

$$x_i^{j,t+1} = x_i^{j,t} + b_w(r_3) \text{ if } HMCR \leq r_1 \text{ and } PAR \leq r_2 \text{ (for continuous variables)} \quad (28)$$

Eq. (25) is the global optimization part of the algorithm and it represents the generation of a totally new solution within a range with a minimum ($x_{i,min}$) and a maximum ($x_{i,max}$). Eq. (26) is the acceptance of the existing value and Eqs. (27) and (28) are local optimization formulation for discrete and continuous design variables, respectively. Three random numbers (r_1, r_2 and r_3) are used in formulations. r_1 and a r_2 are assigned between 0 and 1, while r_3 is assigned between -1 and 1. The generated better results are always accepted by eliminating existing solutions. The essential optimization flowchart of HS is given in Fig. 12.

Jaya Algorithm

Jaya algorithm (JA) is a single-phase algorithm and has no user-defined parameter. In that case, it is an easy user-friendly algorithm. The name "Jaya" comes from the Sanskrit word meaning "Victory". Rao (2016) probably gave that name since finding the best solution after an optimization process is the victory of science. The formulation of the single-phase is shown as Eq. (29).

$$x_i^{j,t+1} = x_i^{j,t} + r_1(x_i^* - |x_i^{j,t}|) - r_2(x_i^w - |x_i^{j,t}|) \text{ for } i=1, 2, \ldots, n \text{ and } j=1, 2, \ldots, p \text{ and } t=1, 2, \ldots, t_{max} \quad (29)$$

As seen in Eq. (29), only two random numbers between 0 and 1 (r_1 and r_2) are used with the best and worst existing solutions. The idea is to converge to best one while diverging to worst solution. For that reason, JA has a good convergence ability. All newly generated solutions are compared with existing ones and better ones in value of objective function are stored. The flowchart of single-phase JA is shown as Fig. 13.

A Basic Structural Optimization Example: Vertical Deflection Minimization Problem of I-Beam

In this chapter, a basic structural optimization problem is presented. The aim of the optimization is to find the design variables minimizing vertical deflection of the beam. This problem is mentioned by Gold and Krishnamurty (1977). The beam with I shaped cross section is subjected to two loads; P and Q as seen in Fig. 14.

The design constants are design loads; P, Q, length of the beam (L) and modulus of elasticity which are taken as 600 kN, 50 kN, 200 cm and 20000 kN/cm^2, respectively. The deflection of beam is written according to Eq. (30).

$$f(x) = \frac{PL^3}{48EI} \quad (30)$$

Figure 12. Flowchart of HS

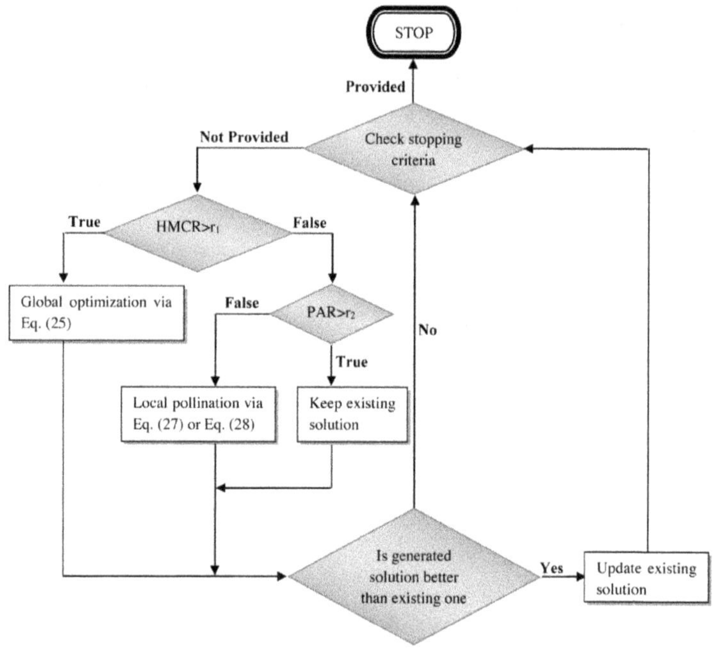

Figure 13. Flowchart of JA

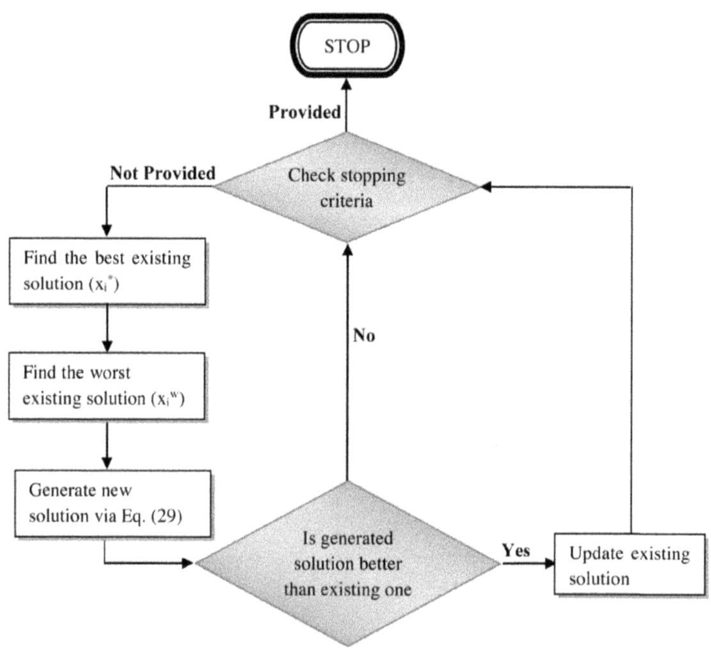

Figure 14. I-beam problem

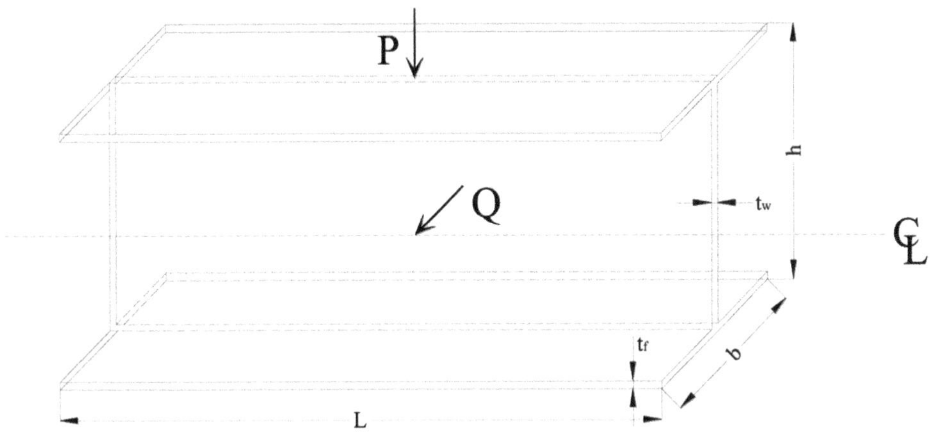

Respect to design variables such as b, h, t_w and t_f, the objective function of the problem is written as Eq.(31) when the formulation of moment of inertia (I) of the I beam and design constants are defined in the equation.

$$Minimize\ f\left(b,h,t_w,t_f\right)=\frac{5000}{\frac{t_w\left(h-2t_f\right)^3}{12}+\frac{bt_f^3}{6}+2bt_f\left(\frac{h-t_f}{2}\right)^2} \quad (31)$$

The ranges of the design variables are as follows:

$10 \leq h \leq 80,$

$10 \leq b \leq 50,$

$0.9 \leq t_w \leq 5$

and

$$0.9 \leq t_f \leq 5. \quad (32)$$

Table 2. Optimum results for I-beam example

Description	h	b	t_w	t_f	$f(b, h, t_w, t_f)$
Result	80.00	50.00	0.9	2.3217922	0.0130741

As design constraints, the cross-section of I-beam must be less than 300 cm^2 for the first one (g_1) and the allowable bending stress of the beam is 6 kN/cm^2 as the second one (g_2). In that situation, cross-section and stress constraints are written as Eqs. (33) and (34).

$$g_1 = 2bt_w + t_w\left(h - 2t_f\right) \leq 300 \tag{33}$$

$$g_2 = \frac{180000h}{t_w\left(h - 2t_f\right)^3 + 2bt_w\left(4t_f^2 + 3h\left(h - 2t_f\right)\right)} + \frac{15000b}{t_w^3\left(h - 2t_f\right) + 2t_w b^3} \leq 6 \tag{34}$$

In the appendix of this chapter, a Matlab Code (The MathWorks Inc., 2018) is presented for application of FPA based method for the problem. By using FPA (Nigdeli, Bekdaş, & Yang, 2016), the optimum design results are given in Table 2.

REFERENCES

Cormen, T. H., Leiserson, C. E., Rivest, R. L., & Stein, C. (2009). *Introduction to algorithms*. MIT Press.

Dorigo, M., Maniezzo, V., & Colorni, A. (1996). The ant system: Optimization by a colony of cooperating agents. *IEEE Transactions on Systems Man and Cybernet B*, *26*(1), 29–41. doi:10.1109/3477.484436 PMID:18263004

Erol, O. K., & Eksin, I. (2006). A new optimization method: Big bang–big crunch. *Advances in Engineering Software*, *37*(2), 106–111. doi:10.1016/j.advengsoft.2005.04.005

Gandomi, A. H., & Alavi, A. H. (2012). Krill herd: A new bio-inspired optimization algorithm. *Communications in Nonlinear Science and Numerical Simulation, 17*(12), 4831–4845. doi:10.1016/j.cnsns.2012.05.010

Geem, Z. W., Kim, J. H., & Loganathan, G. V. (2001). A new heuristic optimization algorithm: Harmony search. *Simulation, 76*(2), 60–68. doi:10.1177/003754970107600201

Glover, F. (1986). Future paths for integer programming and links to artificial intelligence. *Computers & Operations Research, 13*(5), 533–549. doi:10.1016/0305-0548(86)90048-1

Gold, S., & Krishnamurty, S. (1997, September). Trade-offs in robust engineering design. *Proceedings of DETC, 97*, 1997.

Goldberg, D. E., & Samtani, M. P. (1986). Engineering optimization via genetic algorithm. *Proceedings of Ninth Conference on Electronic Computation.* ASCE, New York, NY, pp. 471-482.

Holland, J. H. (1975). *Adaptation in Natural and Artificial Systems.* Ann Arbor, MI: University of Michigan Press.

Karaboga, D., & Basturk, B. (2007). A powerful and efficient algorithm for numerical function optimization: Artificial bee colony (ABC) algorithm. *Journal of Global Optimization, 39*(3), 459–471. doi:10.100710898-007-9149-x

Kaveh, A., & Khayatazad, M. (2012). A new meta-heuristic method: Ray optimization. *Computers & Structures, 112*, 283–294. doi:10.1016/j.compstruc.2012.09.003

Kennedy, J., & Eberhart, R. C. 1995. Particle swarm optimization. In *Proceedings of IEEE International Conference on Neural Networks No. IV*, November 27-December 1, pp. 1942–1948, Perth Australia. 10.1109/ICNN.1995.488968

Kirkpatrick, S., Gelatt, C. D., & Vecchi, M. P. (1983). Optimization by simulated annealing. *Science, 220*(4598), 671–680. doi:10.1126cience.220.4598.671 PMID:17813860

Mirjalili, S., Mirjalili, S. M., & Lewis, A. (2014). Grey wolf optimizer. *Advances in Engineering Software, 69*, 46–61. doi:10.1016/j.advengsoft.2013.12.007

Nigdeli, S. M., Bekdaş, G., & Yang, X. S. (2016). Application of the flower pollination algorithm in structural engineering. In *Metaheuristics and optimization in civil engineering* (pp. 25–42). Cham, Switzerland: Springer. doi:10.1007/978-3-319-26245-1_2

Parpinelli, R. S., & Lopes, H. S. (2011). New inspirations in swarm intelligence: A survey. *International Journal of Bio-inspired Computation*, *3*(1), 1–16. doi:10.1504/IJBIC.2011.038700

Polya, G. (2004). *How to solve it: A new aspect of mathematical method.* Princeton University Press.

Rajpurohit, J., Sharma, T. K., Abraham, A., & Vaishali, A. (2017). Glossary of metaheuristic algorithms. *Int. J. Comput. Inf. Syst. Ind. Manag. Appl*, *9*, 181–205.

Rao, R. (2016). Jaya: A simple and new optimization algorithm for solving constrained and unconstrained optimization problems. *International Journal of Industrial Engineering Computations*, *7*(1), 19–34.

Rao, R. V., Savsani, V. J., & Vakharia, D. P. (2011). Teaching–learning-based optimization: A novel method for constrained mechanical design optimization problems. *Computer Aided Design*, *43*(3), 303–315. doi:10.1016/j.cad.2010.12.015

Simon, H. A., & Newell, A. (1958). Heuristic problem solving: The next advance in operations research. *Operations Research*, *6*(1), 1–10. doi:10.1287/opre.6.1.1

Sörensen, K. (2015). Metaheuristics—The metaphor exposed. *International Transactions in Operational Research*, *22*(1), 3–18. doi:10.1111/itor.12001

Sörensen, K., Sevaux, M., & Glover, F. (2018). A history of metaheuristics. Handbook of heuristics, 1-18.

Storn, R., & Price, K. (1997). Differential evolution–a simple and efficient heuristic for global optimization over continuous spaces. *Journal of Global Optimization*, *11*(4), 341–359. doi:10.1023/A:1008202821328

The MathWorks Inc. (2018). *MATLAB R2018a*. Natick, MA.

Weyland, D. (2010). A rigorous analysis of the harmony search algorithm: How the research community can be misled by a "novel" methodology. [IJAMC]. *International Journal of Applied Metaheuristic Computing*, *1*(2), 50–60. doi:10.4018/jamc.2010040104

Yang, X. S. (2010). A new metaheuristic bat-inspired algorithm. In *Nature inspired cooperative strategies for optimization (NICSO 2010)* (pp. 65–74). Berlin, Germany: Springer. doi:10.1007/978-3-642-12538-6_6

Yang, X. S. (2010). Firefly Algorithm, Stochastic Test Functions and Design Optimisation. *International Journal of Bio-inspired Computation*, *2*(2), 78–84. doi:10.1504/IJBIC.2010.032124

Yang, X. S. (2012). Flower pollination algorithm for global optimization. In *International conference on unconventional computing and natural computation* (pp. 240-249). Berlin, Germany: Springer. 10.1007/978-3-642-32894-7_27

Yang, X. S., & Deb, S. (2010). Engineering Optimisation by Cuckoo Search. *International Journal of Mathematical Modelling and Numerical Optimisation*, *1*(4), 330–343. doi:10.1504/IJMMNO.2010.035430

APPENDIX

```
% clear memory
clear all
%1.ENTER DATA OF PROBLEM
% sigma: allowable bending stress
% E   : modules of elasticity
% P   : design load in vertical direction
% Q   : design load in horizontal direction
% L   : length of the beam
sigma=6; E=20000; P=600; Q=50; L=200;
% User defined algorithm parameters
% maxiter: maximum iteration number
% pn    : population number
% ps    : switch probability
maxiter=20000; pn=15; ps=0.5;
% Ultimate limits of design variables
% hmin: lower limit for height of cross section
% hmax: upper limit for height of cross section
% bmin: lower limit for breadth of cross section
% bmax: upper limit for breadth of cross section
% twmin: lower limit for web thickness of cross section
% twmax: upper limit for web thickness of cross section
% tfmin: lower limit for flange thickness of cross section
% tfmax: upper limit for flange thickness of cross section
hmin=10; hmax=80; bmin=10; bmax=50; twmin=0.9; twmax=5;
tfmin=0.9; tfmax=5;
%2. GENERATING OF INITIAL SOLUTION MATRIX
 % h : height of beam
 % b : breadth of beam
 % tw: web thickness of beam
 % tf: flange thickness of beam
 % rand: a function generating uniformly distributed
pseudorandom numbers in MATLAB
for i=1:pn % During this loop, candidate solution vectors
including design variables are generated up to pn.
           % End of this loop, candidate solution vectors
providing design constraints are stored in a matrix.
    %2.1. DEFINITION OF DESING VARIABLES RANDOMLY
    h=hmin+(hmax-hmin)*rand;
    b=bmin+(bmax-bmin)*rand;
    tw=tfmin+(twmax-twmin)*rand;
    tf=tfmin+(tfmax-tfmin)*rand;
    % 2.2. OBJECTIVE FUNCTION
    C=5000/(((1/12)*tw*(h-2*tf)^3)+(((1/6)*b*tf^3)+(2*b*tf*((h-
tf)/2)^2)));
    % 2.3. DESIGN CONSTRAINTS
```

```
    g1=2*b*tf+tw*(h-2*tf);
    g2=(18000*h)/((tw*(h-2*tf)^3)+(2*b*tf*((4*tf^2)+(3*h*(h-
2*tf)))))+(15000*b)/(((h-2*tf)*tw^3)+(2*tw*b^3));
    % 2.4. PART OF PENALIZATION
    % NOTE: Candidate solution vectors that do not meet design
constraints are penalized with a very large objective function
value.
    % With this penaltization, selection of a solution vector
including improper value or values of design variables is
prevented.
     if g1>300
        C=10^6; % Penalization with high value
    end
    if g2>6
       C=10^6; % Penalization with high value
    end
    % 2.5. INITIAL SOLUTION MATRIX
    % NOTE: End of the each loop, a candidate solutution vector
including randomly generated design variables stored in the
matrix named OPT
    % Thus each vector in matrix is candidate solution vector.
    OPT(1,i)=h;
    OPT(2,i)=b;
    OPT(3,i)=tw;
    OPT(4,i)=tf;
    OPT(5,i)=C;
    OPT(6,i)=g1;
    OPT(7,i)=g2;
end
% 3. GENERATING OF NEW SOLUTION MATRIX VIA FPA
% NOTE: This part is very similar to PART 2. Differently, new
solution variables are generated.
% ultimate limits of variables are controlled.
% All steps in the PART3 are repeated until all iteration is
completed.
for iter=1:maxiter
    % 3.1. SELECTION OF GLOBAL or LOCAL POLLINATION ACCORDING
TO SWITCH PROBABILTY
    % 3.1.1. GLOBAL POLLINATION PHASE
    if rand()<ps
        for i=1:pn
        % 3.1.1.1. DEFINATION OF DESING VARIABLES
       % Values of candidate solution vector providing best/
minimum value of objective function in existing solution set or
matrix is assigned to vector named best:
        [p,r]=min(OPT(5,:));
        best1=OPT(1,r);
        best2=OPT(2,r);
        best3=OPT(3,r);
```

```
        best4=OPT(4,r);
        % New values of design variables are generated
according to regulation of FPA algorithm (see Eq.10)
        levy=(1/(2*pi()))*(rand()^-1.5)*exp(-1/(2*rand()));
        h=OPT(1,i)+levy*(best1-OPT(1,i));
        b=OPT(2,i)+levy*(best2-OPT(2,i));
        tw=OPT(3,i)+levy*(best3-OPT(3,i));
        tf=OPT(4,i)+levy*(best4-OPT(4,i));
        % 3.1.1.2. CONTROL OF ULTIMATE LIMITS
        if h>hmax
           h=hmax;
        end
        if h<hmin
           h=hmin;
        end
        if b>bmax
           b=bmax;
        end
        if b<bmin
           b=bmin;
        end
        if tw>twmax
           tw=twmax;
        end
        if tw<twmin
           tw=twmin;
        end
        if tf>tfmax
           tf=tfmax;
        end
        if tf<tfmin
           tf=tfmin;
        end
        % 3.1.1.3. OBJECTIVE FUNCTION
        C=5000/(((1/12)*tw*(h-
2*tf)^3)+(((1/6)*b*tf^3)+(2*b*tf*((h-tf)/2)^2)));
        % 3.1.1.4. DESIGN CONSTRAINTS
        g1=2*b*tf+tw*(h-2*tf);
        g2=(18000*h)/((tw*(h-
2*tf)^3)+(2*b*tf*((4*tf^2)+(3*h*(h-2*tf)))))+(15000*b)/
(((h-2*tf)*tw^3)+(2*tw*b^3));
        % 3.1.1.5. PART OF PENALIZATION
        if g1>300  % Control of design constraint
           C=10^6; % Penalization with high value
        end
        if g2>6    % Control of design constraint
           C=10^6; % Penalization with high value
        end
        % 3.1.1.6. NEW SOLUTION MATRIX
```

```
        OPT1(1,i)=h;
        OPT1(2,i)=b;
        OPT1(3,i)=tw;
        OPT1(4,i)=tf;
        OPT1(5,i)=C;
        OPT1(6,i)=g1;
        OPT1(7,i)=g2;
        % 3.1.1.7. PART OF COMPARISON
        % Every candidate solution vectors in the new solution
matrix named OPT1 are compared with existing solution vectors
in maxtrix OPT
        % If new solution vectors provide better solution or
minimum cost replaced with existing solution vectors:
        if OPT(5,i)>OPT1(5,i)  % Comparing of values of
objective functions
           OPT(:,i)=OPT1(:,i); % Updating of matrix OPT
        end
        end
    % 3.1.2. LOCAL POLLINATION PHASE  (stiuation of rand()>ps)
    else
        for i=1:pn
            % 3.1.2.1. DEFINATION OF DESING VARIABLES
            % New values of design variables are generated
according to Eq. 11
            h=OPT(1,i)+rand()*(OPT(1,(ceil(rand()*pn)))-
OPT(1,(ceil(rand()*pn))));
            b=OPT(2,i)+rand()*(OPT(2,(ceil(rand()*pn)))-
OPT(2,(ceil(rand()*pn))));
            tw=OPT(3,i)+rand()*(OPT(3,(ceil(rand()*pn)))-
OPT(3,(ceil(rand()*pn))));
            tf=OPT(4,i)+rand()*(OPT(4,(ceil(rand()*pn)))-
OPT(4,(ceil(rand()*pn))));
            % 3.1.2.2. CONTROL OF ULTIMATE LIMITS
            if h>hmax
            h=hmax;
            end
            if h<hmin
            h=hmin;
            end
            if b>bmax
            b=bmax;
            end
            if b<bmin
            b=bmin;
            end
            if tw>twmax
            tw=twmax;
            end
            if tw<twmin
```

```
            tw=twmin;
            end
            if tf>tfmax
            tf=tfmax;
            end
            if tf<tfmin
            tf=tfmin;
            end
            % 3.1.2.3. OBJECTIVE FUNCTION
            C=5000/(((1/12)*tw*(h-
2*tf)^3)+(((1/6)*b*tf^3)+(2*b*tf*((h-tf)/2)^2)));
            % 3.1.2.4. DESIGN CONSTRAINTS
            g1=2*b*tf+tw*(h-2*tf);
            g2=(18000*h)/((tw*(h-
2*tf)^3)+(2*b*tf*((4*tf^2)+(3*h*(h-2*tf)))))+(15000*b)/
(((h-2*tf)*tw^3)+(2*tw*b^3));
            % 3.1.2.5. PART OF PENALIZATION
             if g1>300  % Control of design constraint
                C=10^6; % Penalization with high value
             end
             if g2>6
                C=10^6; % Penalization with high value
             end
            % 3.1.2.6. NEW SOLUTION MATRIX
            OPT1(1,i)=h;
            OPT1(2,i)=b;
            OPT1(3,i)=tw;
            OPT1(4,i)=tf;
            OPT1(5,i)=C;
        end
         % 3.1.2.7. PART OF COMPARISON
         for i=1:pn
             if OPT(5,i)>OPT1(5,i)  % Comparing of values of
objective functions
                OPT(:,i)=OPT1(:,i); % Updating of existing
matrix OPT
             end
         end
    end
end
```

Chapter 3
Optimum Design of Reinforced Concrete Beams:
Optimization of RC Beams

ABSTRACT

The design of reinforced concrete (RC) beams need special conditions to provide a ductile design. In this design, the maximum amount of tensile reinforcement must be limited to singly reinforced design. After the singly reinforced limit, the cost of doubly reinforced RC beam rapidly increases, and it may not be an optimum design. To consider this nonlinear behavior and other rules used in RC structures according to regulations such as ACI 318: Building code requirements for structural concrete and Eurocode 2: Design of concrete structures, an algorithmic and iteration optimization method is needed. In this chapter, two examples are presented, and optimum results are shared for methodologies employing several metaheuristic algorithms. The importance of using metaheuristic algorithms can be seen in this chapter.

INTRODUCTION

For a reinforced concrete (RC) structure, loads on the structures are directed to slabs and the loads on slabs are directed to beams. The loads of beams are directed to columns and then to the foundation of the structure. Finally, all loads are directed to ground. Slab, beam and several parts of foundations are designed according to a member under tensile stresses and ductile behavior

DOI: 10.4018/978-1-7998-2664-4.ch003

is provided by allowing the yielding of reinforcing steel bars (rebar) before fracture of concrete. This ductile behavior can be provided since a portion of the section is generally under tensile stresses. Whereas, this situation cannot be provided since columns are generally under compressive stresses and other rules are needed for structural ductility.

In order to provide ductility condition of beams, the amount of rebar must be limited and thus, moment capacity is also limited. In order to increase moment capacity, rebars in the part of RC sections under compressive stress must be provided.

As explained in Chapter 1, due to existence of design constraint provided in design codes, the optimization of RC member is highly non-linear especially for ductile behavior of RC beams. In that case, several metaheuristic-based optimization methods were proposed for RC beams (Coello, Hernández, & Farrera, 1997; Rafiq & Southcombe, 1998; Govindaraj & Ramasamy, 2005; Akın & Saka, 2010; Fedghouche & Tiliouine, 2012; Bekdaş & Nigdeli, 2013).

In this chapter, two applications of RC beams were given. For the first methodology, design of RC beams was presented according to ACI 318: Building code requirements for structural concrete (2005). The optimum results of several cases were presented according to various metaheuristic algorithms such as flower pollination algorithm (FPA) (Yang, 2012), teaching-learning-based optimization (TLBO) (Rao, Savsani, & Vakharia, 2011) and Jaya algorithm (JA) (Rao, 2016).

As the second methodology, optimum design of T-beams using JA was presented by considering design constraints formulated according to Eurocode 2: Design of concrete structures (2005). Also, Matlab (2018) code for optimization of the numerical example is presented as Appendix 1 with comment lines.

OPTIMUM DESIGN OF RC BEAMS VIA ACI-318

Design Under Flexural Moment

Under flexural effects, the section of RC members is under varying stresses respect to the height (h) of the section where maximum and minimum stresses occur at the end-point fibers. As known, concrete has a small tensile strength and this strength is neglected in flexural design of RC members. In that case, compressive stresses are only considered in the area starting from maximum compressive fiber (minimum stress point if compressive forces are assigned as negative) to neutral axis where there are no stresses. As shown in Fig. 1, compressive stress block of concrete is a parabola because of the stress-strain relationship of concrete and for simplicity, it is generally assumed as rectangular stress block by taking α as 0.85 in ACI-318. In that case, the depth of the compressive block will be shorter. All definitions of symbols given in Fig. 1 are shown in Table 1.

The force balance of compressive stresses is provided by tensile forces resulting from the strain of the rebar. By equalization of resultant forces (C=T), distance from extreme compression fiber to neutral axis is written as Eq. (1).

$$c = \frac{A_s f_y}{\alpha f'_c b_w} \tag{1}$$

Figure 1. A RC cross-section under flexure

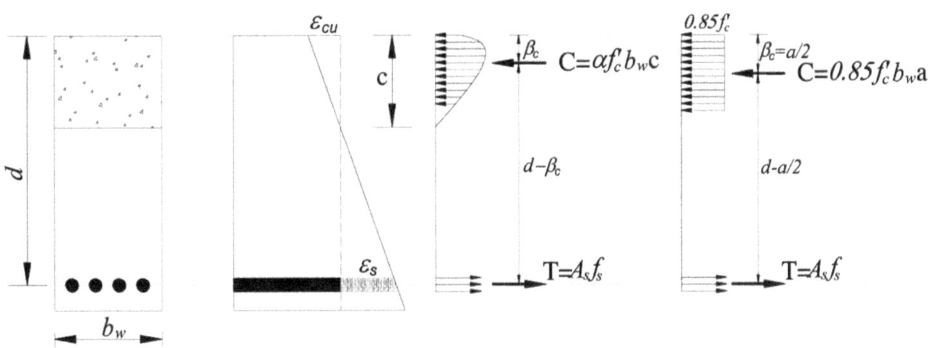

Table 1. Notation and definitions of symbols

b_w = web width, or diameter of circular section
d = distance from extreme compression fiber to centroid of longitudinal tension reinforcement
c = distance from extreme compression fiber to neutral axis
β_c = distance from centroid of compressive stress block to face of compressive section
f'_c = specified compressive strength of concrete
f_s = calculated tensile stress in reinforcement at service loads
A_s = area of nonprestressed longitudinal tension reinforcement
a = depth of equivalent rectangular stress block
ε_s = strain of steel in tensile section
ε_{cu} = strain of concrete in face of compressive section
C = resultant force of compressive block
T = resultant force of tensile reinforcements

The depth of equivalent rectangular stress block is defined as;

$$a = \beta_1 c \tag{2}$$

where β_1 is the factor relating depth of equivalent rectangular compressive stress block to neutral axis depth. This factor is related with ductility and the increase of compressive strength of concrete reduces this factor. In ACI-318, for fc' values between 17 and 28 MPa, β_1 is taken as 0.85. For f_c' above 28 MPa, β_1 is reduced linearly by 0.05 for each 7 MPa of strength increase, but the smallest value to be taken is 0.65.

If the equality of tensile and compressive stresses is done for rectangular stress block; depth of equivalent rectangular stress block is formulated as follows:

$$a = \frac{A_s f_y}{0.85 f'_c b_w} \tag{3}$$

The moment capacity of RC member is written according to Eq. (4) or Eq. (5) by taking moments of C and T respect to moment arm (d- β_c or d-a/2);

$$M = A_s f_s \left(d - \frac{a}{2} \right) \tag{4}$$

$$M = 0.85 f'_c b_w a \left(d - \frac{a}{2} \right) \tag{5}$$

According to experimental results, concrete is crushed after when ε_{cu} is equal 0.003. If concrete is crushed before the yielding of steel, failure is sudden. In balanced situation, steel yields and concrete crushes at the same time. This situation is also brittle fracture. For ductile behavior, concrete must not crush. For that reason, the required reinforcement ratio under balanced situation must be reduced by a factor. This factor is 0.75 in ACI-318 as seen in maximum ratio of A_s to $b_w d$ (ρ_{max}) formulation given in Eq. (6).

$$\rho_{max} = (0.75)(0.85)\beta_1 \frac{f'_c}{f_y}\left(\frac{600}{600+f_y}\right) \quad (6)$$

In this formulation, f_y is specified yield strength of reinforcement. If the calculated maximum reinforcement area does not provide nominal flexural moment (M_n), doubly reinforced design is done to increase compressive forces by addition rebars to compressive fiber. In ACI-318, nominal flexural moment (M_n) is factored with ϕ in order to find factored flexural moment; M_u. For tension-controlled section such as beams, ϕ is 0.9. Additional to the limitation of reinforcements, minimum reinforcement area ($A_{s,min}$) must be also considered according to unfavorable one of Eqs. (7) and (8).

$$A_{s,min} \geq \frac{\sqrt{f'_c}}{4f_y} b_w d \quad (7)$$

$$A_{s,min} \geq \frac{1.4}{f_y} b_w d \quad (8)$$

Design Under Shear Force

To provide capacity for shear stresses, tensile strength of concrete is used, but additional shear reinforcements; namely stirrups may be needed for beams. Nominal shear force (V_u) given as Eq. (9) is provided with nominal shear strength of concrete (V_c) as seen in Eq. (10) and nominal shear strength provided by shear reinforcement (V_s) as seen in Eq. (11). The factored shear force capacity (V_u) is provided by multiplying nominal shear force with a reduction factor.

$$V_u = \varphi V_n \quad (9)$$

$$V_c = \frac{\sqrt{f'_c}}{6} b_w d \quad (10)$$

$$V_s = \frac{A_v f_y d}{s} \tag{11}$$

For ductile design, the allowed shear force must be less than $0.66\sqrt{f'_c}b_w d$ and minimum shear reinforcements must be provided even nominal shear strength of concrete (V_c) is enough in calculations. A_v is the area of shear reinforcement spacing s and minimum allowed value of A_v is given in Eq (12).

$$\left(A_v\right)_{\min} = \frac{1}{3}\frac{b_w s}{f_y} \tag{12}$$

Also, the maximum spacing between shear reinforcement (s_{max}) must be less than d/2. If V_s value is bigger than $0.33\sqrt{f'_c}b_w d$, s_{max} must be taken as d/4.

The Optimization Problem

The optimum design of rectangular RC beam under flexural moment was investigated for factored moments between 50 kNm and 500 kNm according to rules of ACI318. The objective of this optimization problem is the minimization of the material cost of the beam per unit meter as seen in Eq. (13), where C_c, C_s, V_c and W_s are cost of concrete per m^3, cost of reinforced steel per ton, total concrete volume and total weight of reinforced steel, respectively. In other words, specified design variables will be investigated for design with minimum cost. Investigated design variables (shown in Fig. 2) are breadth (b_w), height (h), number (n_1-n_4) and size (ϕ_1- ϕ_4) of the reinforcements in two lines of compressive and tensile sections.

$$\min f(X) = C_c V_c + C_s W_s \tag{13}$$

When the design variables are investigated, specified design variable ranges are utilized and these ranges are given in Table 2. In optimization, discrete dimensions were used for practical production in construction yard. Dimension variables with 50 mm differences were taken. Also, rebar sizes were assigned by 2 mm differences, because these sizes can be found in markets. Additionally, design constants, which are unchanging variables in

Figure 2. The optimized design variables

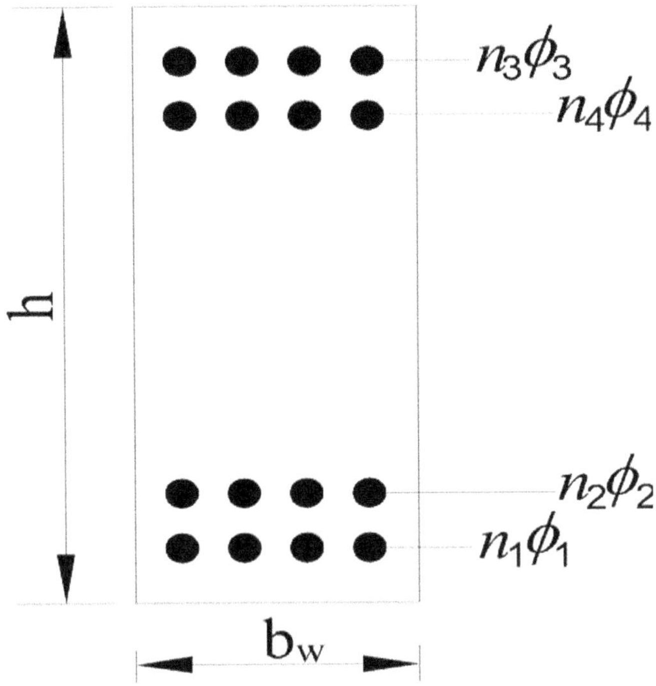

optimization process, are summarized in the Table 3. The cost of steel per ton is taken as 10 times of the cost of concrete per m^3 and a specific money unit is not used.

In optimum design, dimensions are firstly assigned with candidate design variables. Then, the maximum singly reinforced moment capacity is calculated according to maximum reinforcement ratio given as Eq. (7). If the maximum moment capacity is lower than the objective flexural moment value, doubly reinforced design is done by assigning candidate solutions for n3, n4, ϕ_3 and ϕ_4 in addition to tensile reinforcements with variables; n_1, n_2, ϕ_1 and ϕ_2. The placement of reinforcements is checked, and double line reinforcements are

Table 2. Design variable ranges

diameters of the main reinforcement bars (ϕ_1- ϕ_4)	10-30 mm
breadth (b_w)	250-350 mm
height (h)	350-500 mm

Table 3. Design constants

cover of the reinforcement	35 mm
the maximum size of the aggregate diameter	16 mm
the specified compressive strength of concrete	20 MPa
the specified yield strength of reinforcement	420 MPa
the diameter of stirrup	10 mm
the cost of the concrete	40 unit
the cost of the reinforcement bars	400 unit

used. For adherence, the clear distance between bars must be more than the maximum diameter size, 25 mm and 4/3 times of the size of aggregate used in production of concrete.

The Optimum Results

For the several flexural moment cases, several optimum results found by using JA (Rao, 2016), TLBO (Rao et al., 2011) and FPA (Yang, 2012) are presented in Table 4. If the flexural moment is 250 kNm or more, doubly reinforced design is needed. For the numerical cases, robustness evaluation is done by repeating the same optimization cases 20 times. In general, JA is a little better at minimizing the objective function. For different flexural moments, different algorithms may show the best performance and it is the proof of the no-free-lunch theorem. As seen from the small standard deviation results, the algorithms are generally robust. The maximum standard derivative value is 0.43 for 400 kNm flexural moments by using JA algorithm. This situation may be resulted from the single phase of JA.

OPTIMUM DESIGN OF RC T-BEAMS VIA EUROCODE 2

In the second example of RC beams, T-shaped cross section (Fig. 3) is considered for optimization via Eurocode 2: Design of structures (EN 1992-1-1). In the presentation of this example, the design constraints are formulated via Eurocode 2 and five inequality functions (g_i for i=1, 2,..., 5) are given as Eqs. (14)-(18).

Figure 3. T-shaped cross section

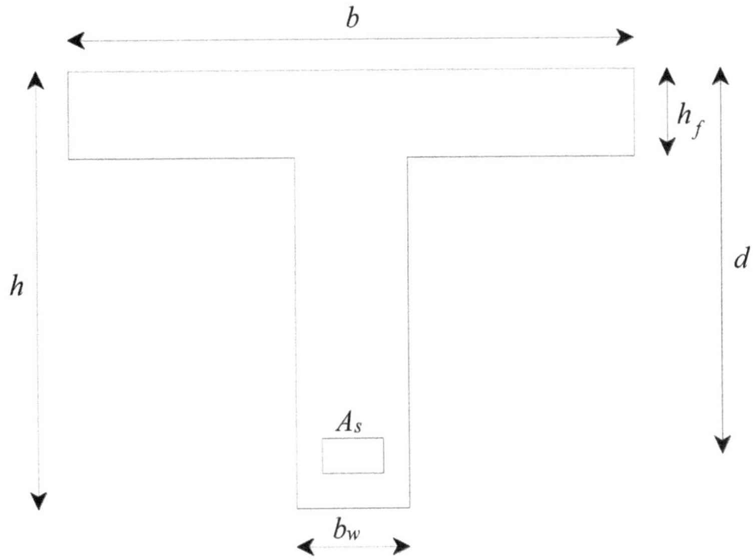

This problem was firstly presented by Fedghouche and Tiliouine (2012) and seven design variables are used in the example. Two of the design variables; effective depth and cover of reinforcement are variables of another design variable; height (h). For that reason, five independent design variables. The ranges and formulations of seven design variables are shown in Table 5. The design constants are given in Table 6 with definitions.

$$g_1 = -\omega\left(1 - 0.5\omega\right) + 0.392 \geq 0 \quad (14)$$

$$g_2 = \frac{0.0035\left(0.8 - \omega\right)}{\omega} - f_{yd} / E_s \geq 0 \quad (15)$$

$$g_3 = \rho_{\min} \leq \rho \leq \rho_{\max} \quad (16)$$

$$g_4 = -M_{Ed} + M_{Ed1} \geq 0 \quad (17)$$

Table 4. The optimum results

JA										
Moment (kNm)	50	100	150	200	250	300	350	400	450	500
h (mm)	350	400	500	500	500	500	500	500	500	500
b_w (mm)	250	250	250	250	250	300	300	350	350	400
ϕ_1 (mm)	14	16	16	28	26	22	26	26	28	26
ϕ_3 (mm)	20	28	30	22	14	12	16	12	16	12
n_1	2	3	4	2	3	5	4	5	5	6
n_3	0	0	0	0	2	2	3	6	5	9
ϕ_2 (mm)	12	10	12	10	12	14	12	10	10	12
ϕ_4 (mm)	18	16	14	16	18	22	18	26	14	30
n_2	2	4	2	3	2	2	4	3	2	4
n_4	0	0	0	0	0	0	0	0	0	0
M_u (kNm)	57.31	111.22	167.71	222.42	279.29	333.48	388.94	445.15	500.11	555.96
Cost (unit/m) (best)	5.16	6.85	8.20	9.55	11.60	13.56	15.87	18.08	20.16	22.45
Num. of analyses	225	100	75	100	125	200	1475	950	175	2100
Cost (unit/m) (ave.)	5.17	6.86	8.25	9.59	11.65	13.70	16.05	18.62	20.58	22.93
Standard Dev.	0.01	0.02	0.07	0.07	0.06	0.12	0.19	0.43	0.28	0.21
TLBO										
Moment (kNm)	50	100	150	200	250	300	350	400	450	500
h (mm)	350	400	500	500	500	500	500	500	500	500
b_w (mm)	250	250	250	250	250	250	300	300	350	400
ϕ_1 (mm)	14	16	16	28	26	28	26	28	28	26
ϕ_3 (mm)	30	30	16	20	14	14	14	16	16	12
n_1	2	3	4	2	3	3	4	4	5	6
n_3	0	0	0	0	2	4	4	5	5	9
ϕ_2 (mm)	12	10	12	10	12	12	14	12	12	12
ϕ_4 (mm)	28	14	16	10	18	18	18	12	16	20
n_2	2	4	2	3	2	3	3	4	2	4
n_4	0	0	0	0	0	0	0	0	0	0
M_u (kNm)	57.31	111.22	167.71	222.42	279.29	333.50	390.51	445.40	506.81	555.96
Cost (unit/m) (best)	5.16	6.85	8.20	9.55	11.60	13.70	15.94	18.17	20.38	22.45

Table 5. Design variables and ranges

Symbol	Definition	Range
b	Effective with of compressive flange [mm]	$b_w \le b \le \min\left[0.2L + b_w, 8h_f\right]$
b_w	Web width [mm]	$0.20d \le b_w \le 0.40d$
h	Height [mm]	$L/16 \le h \le 2.0$
h_f	Flange depth [mm]	$0.15 \le h_f \le d$
d	Effective depth [mm]	$d = 0.9h$
d_s	Cover of reinforcements [mm]	$d_s = 0.1h$
A_s	Area of reinforcing steel [mm²]	$0 \le A_s \le 0.1$

Table 6. Design constants and definitions

Symbol	Definition
f_{ck}	Characteristic compressive strength for concrete
f_{cd}	Allowable compressive strength for concrete
f_{yd}	Characteristic yield strength of reinforcement
ρ_{max}	The maximum reinforcement ratio
ρ_{min}	The minimum reinforcement ratio
L	The length of beam
E_s	Young's elastic modules for steel
M_{Ed}	The ultimate bending moment capacity
V_{Ed}	The ultimate bending moment capacity
C_s	The unit total cost of reinforcing steel
C_c	The unit total cost of concrete

$$g_5 = -V_{Ed} + V_{Rd\max} \ge 0 \tag{18}$$

The constraints; Eqs. (14) and (15) are related to the ductile behavior of beam. Eq. (16) represents the minimum and maximum allowed reinforcement ratio in Eurocode 2 design. Eq. (17) and (18) represent the flexural moment and shear force capacity of beam. In optimum design, only singly reinforced design is allowed. For the other values given in formulation of design constraints, the expressions required in the calculation are as follows:

$$\omega = \left(f_{yd} / f_{cd}\right)\left(A_s / b_w d\right) - \left(b - b_w\right)h_f / \left(b_w d\right) \tag{19}$$

$$\rho = A_s / \left(b_w d\right) \tag{20}$$

$$M_{Ed1} = f_{cd}\left(b - b_w\right)h_f\left(d - 0.50h_f\right) + f_{cd}b_w d^2\omega\left(1 - 0.5\omega\right) \tag{21}$$

$$V_{Rd\max} = \nu_1 f_{cd} b_w z / \left(\tan\left(45\right) + \cot\left(45\right)\right) \tag{22}$$

In the Eq. (22), ν_1 and z are non-dimensional coefficient and lever arm as given in Eqs. (23) and (24).

$$\nu_1 = 0.6(1 - f_{ck} / 250) \tag{23}$$

$$z = 0.9d \tag{24}$$

The objective function used in the minimization is the material cost of the beam as formulated in Eq. (25)

$$C = b_w d + \left(b - b_w\right)h_f + \left(C_s / C_c\right)A_s \tag{25}$$

The Optimum Results

The presented optimization problem was solved via Genetic Algorithm (GA) by Fedghouche and Tilioune (2012) and via Jaya algorithm (JA) by Kayabekir, Bekdaş and Nigdeli (2019). The Matlab code using JA is also presented in the appendix of this section with comment line descriptions. The numerical values used in the optimization are presented in Table 7.

The optimum design variables are given in Table 8 by using a ratio for the cost objective function. Also, a classical solution done by hand calculations is presented. As clearly seen, the optimization has 10.5% economy on the total material cost. JA is slightly better in minimization of total cost. As the discussion of results, user friendly and single-phase JA can outperform a well-known classical algorithm.

Table 7. Design constants of numerical example

Symbol	Values
f_{ck}	20 MPa
f_{cd}	11.33 MPa
f_{yd}	348 MPa
ρ_{max}	0.04
ρ_{min}	0.0013
L	20 m
E_s	200000 MPa
M_{Ed}	4.991 N.m
V_{Ed}	1.039 N
C_s/C_c	36

Table 8. Optimum design variables

Design Variables	Classical Solution	Optimum Results (GA)	Optimum Results (JA)
b (m)	1.2	1.2	1.137450
b_w (m)	0.4	0.30	0.304476
h (m)	1.6	1.67	1.691535
h_f (m)	1.46	1.50	1.522382
d (m)	0.14	0.15	0.150009
A_s (m^2)	0.011702	0.01143	0.011404
ω	0.424	0.477	0.486112
C/C_c	1.117272	0.999221	0.999036

REFERENCES

ACI-318. (2005). Building code requirements for structural concrete and commentary, metric version, American Concrete Institute.

Akin, A., & Saka, M. P. (2010), Optimum Detailed Design of Reinforced Concrete Continuous Beams using the Harmony Search Algorithm, In The Tenth International Conference on Computational Structures Technology, Paper 131, Stirlingshire, UK. Academic Press.

Bekdaş, G., & Nigdeli, S. M. (2013). Optimization of T-shaped RC flexural members for different compressive strengths of concrete. *International Journal of Mechanics*, *7*, 109–119.

Bekdaş, G., & Nigdeli, S. M. (2018, May). Robustness of Metaheuristic Algorithms in Optimum Design of Reinforced Concrete Beams, *International Conference on Bioinspired Optimization Methods and their Applications (BIOMA 2018)* (pp. 158-160).

Coello, C. C., Hernández, F. S., & Farrera, F. A. (1997). Optimal design of reinforced concrete beams using genetic algorithms. *Expert Systems with Applications*, *12*(1), 101–108. doi:10.1016/S0957-4174(96)00084-X

EN. (2005). *(Veranst.): EN 1992-1-1 Eurocode 2: Design of concrete structures.* Brussels, Belgium: CEN.

Fedghouche, F., & Tiliouine, B. (2012). Minimum cost design of reinforced concrete T-beams at ultimate loads using Eurocode2. *Engineering Structures*, *42*, 43–50. doi:10.1016/j.engstruct.2012.04.008

Govindaraj, V., & Ramasamy, J. V. (2005). Optimum detailed design of reinforced concrete continuous beams using genetic algorithms. *Computers & Structures*, *84*(1-2), 34–48. doi:10.1016/j.compstruc.2005.09.001

Kayabekir, A. E., Bekdaş, G., & Nigdeli, S. M. (2019, April). Optimum Design of T-Beams Using Jaya Algorithm. In *Proceedings 3rd International Conference on Engineering Technology and Innovation (ICETI)*, (pp. 13).

Mathworks. (2018). MATLAB R2018a. The MathWorks Inc., Natick, MA.

Rafiq, M. Y., & Southcombe, C. (1998). Genetic algorithms in optimal design and detailing of reinforced concrete biaxial columns supported by a declarative approach for capacity checking. *Computers & Structures*, *69*(4), 443–457. doi:10.1016/S0045-7949(98)00108-4

Rao, R. (2016). Jaya: A simple and new optimization algorithm for solving constrained and unconstrained optimization problems. *International Journal of Industrial Engineering Computations*, *7*(1), 19–34.

Rao, R. V., Savsani, V. J., & Vakharia, D. P. (2011). Teaching–learning-based optimization: A novel method for constrained mechanical design optimization problems. *Computer Aided Design*, *43*(3), 303–315. doi:10.1016/j.cad.2010.12.015

Yang, X. S. (2012). Flower pollination algorithm for global optimization. In *International conference on unconventional computing and natural computation* (pp. 240-249). Berlin, Germany: Springer. doi:10.1007/978-3-642-32894-7_27

APPENDIX

```
clear all
clc
%1.ENTER DATA OF PROBLEM
fck=20;
fyd=348;
Es=200000;
fcd=11.33;
pmax=0.04; %In accordance with EUROCODE 2(EC2)
pmin=0.0013; %In accordance with EC2
L=20;
MEd=4.991;
VEd=1.039;
Cs=36;
Cc=1;
Cf=0;
pn=10; % pn is population number defined by user
maxiter=100000; %maximum iteration number defined by user
%2. GENERATING OF INITIAL SOLUTION MATRIX
for i=1:pn % During the this loop, candidate solution vectors
including design variables are generated up to pn.
           % End of the this loop, candidate solution vectors
provided design constraint are stored in a matrix.
    %2.1. DEFINITION OF DESING VARIABLES
    %NOTE: Design variables and its limits can be seen in Table
5.
    %NOTE: Mentioned design variables are going to be between
its ultimate limits and generated randomly.
    C=10^6; %C is value of objective function and 10^6 is
penalization value.

    while C==10^6%With this compute, design variables which are
proper to design constraints are obtained only.
        hmin=L/16;                  %hmin is lower bound of h
        hmax=2.0;                   %hmax is upper bound of h
        h=hmin+rand*(hmax-hmin); %Generation of h between
ultimate limits and randomly
        d=0.9*h;  %d was generated according to h
        ds=0.1*h; %ds was generated according to h
        bwmin=0.2*d;                    %bwmin is lower bound of
bw
        bwmax=0.4*d;                    %bwmax is upper bound of
bw
        bw=bwmin+rand*(bwmax-bwmin); %Generation of bw randomly
        hfmin=0.15;                     %hfmin is lower bound of
hf
```

```
        hfmax=d;                        %hfmax is upper bound of
hf
        hf=hfmin+rand*(hfmax-hfmin); %Generation of hf randomly
        bmin=bw;                     %bmin is lower bound of b
        bmax=min(0.2*L+bw,8*hf);  %bmax is upper bound of b
        b=bmin+rand*(bmax-bmin);  %Generation of b randomly
        Asmin=0;                        %As is lower bound of As
        Asmax=0.1;                      %As is upper bound of As
        As=Asmin+rand*(Asmax-Asmin); %Generation of As randomly
    % 2.2. OTHER CALCULATIONS ACCORDING TO EC2
        w=(fyd/fcd)*(As/(bw*d))-(b-bw)*hf/(bw*d);               %
This function was defined with Eq.(19)
        p=As/(bw*d);                                            %
The reinforcement ratio was defined with Eq.(20)
        if p<pmin
           p=pmin;
        end
        MEd1=fcd*(b-bw)*hf*(d-0.5*hf)+fcd*bw*d^2*w*(1-0.5*w); %
This function was defined with Eq.(21)
        v1=0.6*(1-fck/250);                                     %
The non-dimensional coefficient was defined with Eq.(23)
        z=0.9*d;                                                %
The lever arm was defined with Eq.(24)
        VRdmax=v1*fcd*bw*z/(tand(45)+cotd(45));                 %
This function was defined with Eq.(22)
    % 2.3. OBJECTIVE FUNCTON
    %NOTE: The values of objective function is calculated in
every cycle and through minimum value of objective function
    %the  candidate solution vector including values of
candidate optimum design variables can be found .
        C=bw*d+(b-bw)*hf+(Cs/Cc)*As+Cf/Cc*(b+2*(ds+d));
    % 2.4. PART OF PENALIZATION
    %NOTE: In this part,if generated any design variables do not
provided constraint, objective function are penalized with a
large value.
    %With this penaltization, selection of a solution vector
including improper values of generated design variable is
prevented.
        if w*(1-0.5*w)>0.392          % Control of design
constraint according to Eq. (14)
           C=10^6;                    % Penalization with high
value
        end
        if 0.0035*(0.8-w)/w<fyd/Es  % Control of design
constraint according to Eq. (15)
           C=10^6;                    % Penalization with high
value
        end
        if p>pmax                     % Control of design
```

```
constraint according to Eq. (16)
           C=10^6;                  % Penalization with high
value
        end
        if MEd>MEd1                 % Control of design
constraint according to Eq. (17)
           C=10^6;                  % Penalization with high
value
        end
         if VEd>VRdmax              % Control of design
constraint according to Eq. (18)
            C=10^6;                 % Penalization with high
value
        end
    end
   % 2.5. INITIAL SOLUTION MATRIX
   %NOTE: In this part, all the design variables correspounding
to minimum value of objetive function
   %and value of objective function were stored in a matrix
named OPT.
   %Every column of this matrix is a candidate solution vector.
    OPT(1,i)=b;
    OPT(2,i)=bw;
    OPT(3,i)=h;
    OPT(4,i)=d;
    OPT(5,i)=hf;
    OPT(6,i)=As;
    OPT(7,i)=w;
    OPT(8,i)=C;
end
% 3. GENERATING OF NEW SOLUTION MATRIX VIA JA
% NOTE: This part is very similar to PART 2. Differently, new
solution variables generated acoording to JA.
% and ultimate limits of variables are controlled.
% All steps during the PART3 are repeated until all iteration
is completed.
for iter=1:maxiter
    for i=1:pn
    % 3.1. DEFINATION OF DESING VARIABLES
        % Values of candidate solution vector providing best/
minimum value of objective function in existing solution set or
matrix are assigned to vector named best:
        [pe,re]=min(OPT(8,:));
        best1=OPT(1,re);
        best2=OPT(2,re);
        best3=OPT(3,re);
        best4=OPT(4,re);
        best5=OPT(5,re);
        best6=OPT(6,re);
```

```
        % Values of candidate solution vector providing worst/
maximum value of objective function in existing solution set or
matrix are assigned to vector named worst:
        [ke,te]=max(OPT(8,:));
        worst1=OPT(1,te);
        worst2=OPT(2,te);
        worst3=OPT(3,te);
        worst4=OPT(4,te);
        worst5=OPT(5,te);
        worst6=OPT(6,te);
        % New values of design variables are generated
according to JA
        b=OPT(1,i)+rand*(best1-abs(OPT(1,i)))-rand*(worst1-
abs(OPT(1,i)));
        bw=OPT(2,i)+rand*(best2-abs(OPT(2,i)))-rand*(worst2-
abs(OPT(2,i)));
        h=OPT(3,i)+rand*(best3-abs(OPT(3,i)))-rand*(worst3-
abs(OPT(3,i)));
        hf=OPT(5,i)+rand*(best5-abs(OPT(5,i)))-rand*(worst5-
abs(OPT(5,i)));
        As=OPT(6,i)+rand*(best6-abs(OPT(6,i)))-rand*(worst6-
abs(OPT(6,i)));
        d=0.9*h;
        ds=0.1*h;
    % 3.2. CONTROL OF ULTIMATE LIMITS
        if h<L/16
           h=L/16;
        end
        if h>2
           h=2;
        end
        if bw<0.2*d
           bw=0.2*d;
        end
        if bw>0.4*d
           bw=0.4*d;
        end
        if hf<0.15
           hf=0.15;
        end
        if hf>d
           hf=d;
        end
        if b>0.2*L+bw
           b=0.2*L+bw;
        end
        if b>8*hf
           b=8*hf;
        end
```

```
        if b<bw
           b=bw;
        end
    % 3.3. OTHER CALCULATIONS
        w=(fyd/fcd)*(As/(bw*d))-(b-bw)*hf/(bw*d);
        p=As/(bw*d);
        if p<pmin
           p=pmin;
        end
        MEd1=fcd*(b-bw)*hf*(d-0.5*hf)+fcd*bw*d^2*w*(1-0.5*w);
        v1=0.6*(1-fck/250);
        z=0.9*d;
        VRdmax=v1*fcd*bw*z/(tand(45)+cotd(45));
    % 3.4. OBJECTIVE FUVCTION
        C=bw*d+(b-bw)*hf+(Cs/Cc)*As+Cf/Cc*(b+2*(ds+d));
    % 3.5. PART OF PENALIZATION
   %NOTE: In this part,if generated any design variables do not
provided constraint, objective function are penalized with a
large value.
   %With this penaltization, selection of a solution vector
including improper values of generated design variable is
prevented.
        if w*(1-0.5*w)>0.392        % Control of design
constraint according to Eq. (14)
           C=10^6;                  % Penalization with high
value
        end
        if 0.0035*(0.8-w)/w<fyd/Es  % Control of design
constraint according to Eq. (15)
           C=10^6;                  % Penalization with high
value
        end
        if p>pmax                   % Control of design
constraint according to Eq. (16)
           C=10^6;                  % Penalization with high
value
        end
        if MEd>MEd1                 % Control of design
constraint according to Eq. (17)
           C=10^6;                  % Penalization with high
value
        end
         if VEd>VRdmax              % Control of design
constraint according to Eq. (18)
           C=10^6;                  % Penalization with high
value
        end
    % 3.6. NEW SOLUTION MATRIX
     OPT1(1,i)=b;
```

```
     OPT1(2,i)=bw;
     OPT1(3,i)=h;
     OPT1(4,i)=d;
     OPT1(5,i)=hf;
     OPT1(6,i)=As;
     OPT1(7,i)=w;
     OPT1(8,i)=C;
    end
    % 3.7. PART OF COMPARISON
    % NOTE: In this part, each of new candidate solution
vectors in the OPT1  matrix is compared with old ones.
    % If any new solution vector have a better value of
objective function than old one, new values are replaced with
old values.
    % In this way, The matrix named OPT is updated in the every
iteration.
    for i=1:pn
        if OPT(8,i)>OPT1(8,i)
           OPT(:,i)=OPT1(:,i);
        end
    end
%NOTE: The minimum values of objective function can be seen in
command window for every iteration via last two comutes.
[mi,ni]=min(OPT(8,:));
mi
end
```

Chapter 4
Optimum Design of Reinforced Concrete Columns:
Optimization of RC Columns

ABSTRACT

In the design of reinforced concrete (RC) columns, ductility is provided by allowing yielding of steel in the part of section under tensile stresses. This situation cannot be provided for RC columns since sections of columns are generally under compressive stresses resulting from axial loading including weight of all upper stories, flexural moments, and shear forces. To practically provide ductility, axial force is limited, and stirrups are densely designed. These rules are given in design regulations and must be checked during optimization. In this chapter, an optimum design methodology for biaxial loaded column is presented. Uniaxial loaded column methodology is given with the computer code. Finally, the slenderness effects are presented via ACI 318: Building code requirements for structural concrete and optimum results are given for several numerical cases using various metaheuristic algorithms.

INTRODUCTION

Columns of structures are the most important part and design of columns must be provided in the safest rules in design. In structures, all vertical loads are supported by columns. The major parts of vertical loads are self-weight of RC members and all loads on stories above columns are directed via slabs

DOI: 10.4018/978-1-7998-2664-4.ch004

to beams, and beams to columns as axial forces. Since axial forces are big, flexural moments are not generally effective at changing stresses as tensile stresses in sections. For that reason, reinforced concrete (RC) columns are compressive controlled members and the rules for these members are considered in design since the yielding of rebar cannot be provided before concrete and crushing of concrete is not acceptable for columns. In order to ensure ductility, the general rules in design codes are as follows:

- Use a limitation of axial force.
- Use dense spiral reinforcement or stirrups in critical sections.

Also, second order effects are needed to be considered for columns since axial forces are big and horizontal deflection under dynamic forces appears.

In this chapter, an optimum design of biaxial loaded column with flexural moments around two directions is presented by using a modified harmony search (HS) approach. Then, a method using JA and uniaxial loaded columns is given with the optimum design code. Finally, the design formulations to consider slenderness effects are given and the optimum results are presented for different algorithms.

OPTIMUM DESIGN OF RC BIAXIAL LOAD COLUMNS

In this section, the design methodology using a modified harmony search (HS) algorithm developed by Nigdeli, Bekdaş, Kim and Geem (2015) is presented. In this methodology, classical HS (Geem, Kim, & Loganathan, 2001) is combined with several random search stages. These random search stages are used for the two following reasons:

- If a set of design variables do not provide a design constraint, are directly neglected after a violation and a new set is generated. This process continues until all constraints are provided.
- In the optimum design, there are several design variables that are related to each other. Also, the required safety criteria are respected to various types of loads such as flexural moment, axial force and shear force. The optimum design must be suitable for multiple types of loadings.

To conduct a full iterative stage, the number of variable evaluations may increase. In that case, computational time is saved by neglecting violated solutions via additional random stages.

The Optimization Problem

In the optimum design, RC column is subjected to a shear force (V_x), an axial force (N_z) and two flexural moments (M_x and M_y) as seen in Fig. 1.

Figure 1. RC biaxial loaded column

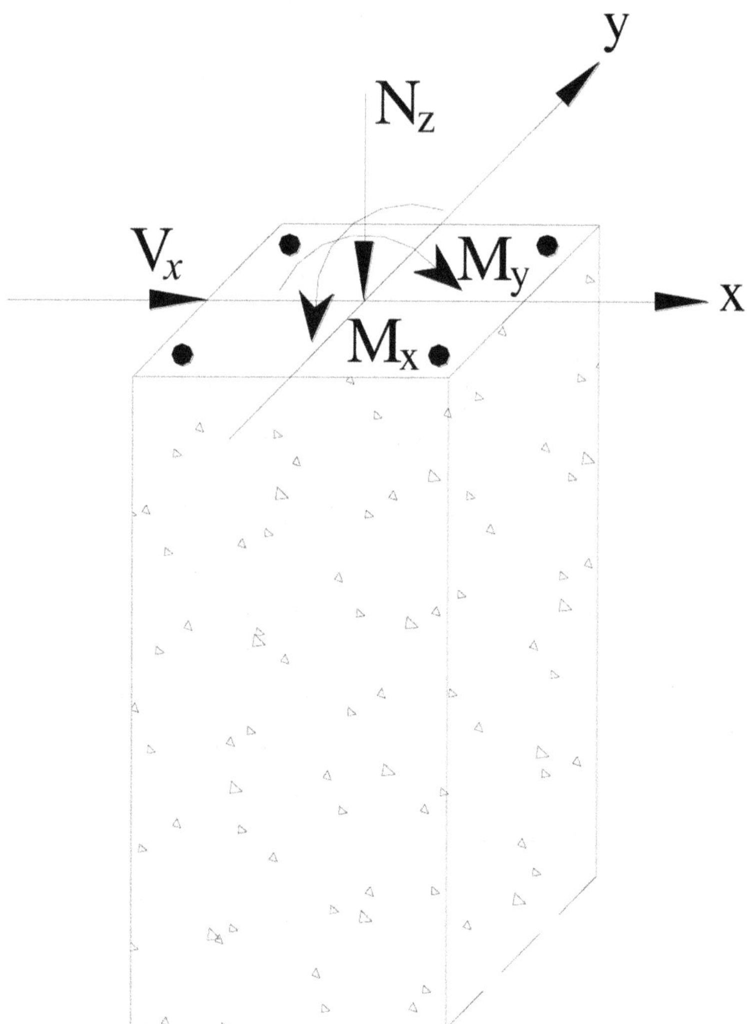

Firstly, the design constants given as Table 1, design variable ranges and algorithm-specific parameters such as harmony memory size (HMS), harmony memory considering rate (HMCR) and pitch adjusting rate (PAR) are defined. HMS is the number of sets of candidate design variables used in the harmony memory (HM) matrix. HMCR is rate of usage of local search instead of global search using all search domains within the selected range. In local search, a new design variable is generated by choosing a random existing variable and value of new variable will be a neighborhood value of it.

The design is done via ACI 318: Building code requirements for structural concrete and design constraints are formulated according to it. The formulation of elasticity modulus of concrete (E_c) is as follows:

$$E_c = 4700\sqrt{f_c'} \tag{1}$$

Then, an initial HM matrix is generated and candidate design variables (x_i for i[th] design variable) are chosen according to Eq. (2) by using minimum ($x_{i,min}$) and maximum ($x_{i,max}$) bounds of related design variables with a random number between 0 and 1 (rand (0,1).

$$x_i = x_{i,\min} + (x_{i,\max} - x_{i,\min})rand(0,1) \tag{2}$$

Table 1. Design constants of the optimization

Definition	Symbol
Flexural moment in both directions	M_{dx} and M_{dy}
Shear force	V_d
Axial force	N_d
Length of column	l
Strain corresponding ultimate stress of concrete	ε_c
Max. aggregate diameter	D_{max}
Yield strength of steel	f_y
Compressive strength of concrete	
Elasticity modulus of steel	E_s
Specific gravity of steel	γ_s
Specific gravity of concrete	γ_c
Cost of the concrete per m³	C_c
Cost of the steel per ton	C_s

The design variables are defined as Eq. (2) as seen in a HM vector. These vectors generated HM matrix. The first two design variables are related to the dimension of cross-section. In Eq. (3), n represents number of longitudinal reinforcements and x_3-x_{n+2} denotes candidate values of longitudinal reinforcement sizes. x_{n+2} denotes the size of shear reinforcement spacing with x_{n+4}.

In generation of candidate solutions, x_1 and x_2 (b and h) which are dimension design variables are firstly assigned. Then, two design constraints such as $g_1(x)$ and $g_2(x)$ are checked because these two constraints are not related with rebar. The first two constraints are related with ductile behavior and shear force and axial force capacities are limited. All design constraints are given in Table 2.

$$X = \begin{bmatrix} x_1 : breadth\,of\,column\left(b\right) \\ x_2 : height\,of\,column\left(h\right) \\ x_3 - x_{n+2} : size\,of\,longitudinal\,reinforcements \\ x_{n+3} : size\,of\,shear\,reinforcements \\ x_{n+4} : distance\,between\,shear\,reinforcements \end{bmatrix} \quad (3)$$

The generation of dimension design variables continues until the first two constraints are provided. Then, the number of longitudinal reinforcements is defined a random number between 2 and the maximum allowed number of reinforcements. The maximum allowed number of reinforcements is calculated according to the third constraint. This constraint defines the limitation of maximum clear distance between longitudinal rebar (a_ϕ).

Generally, a symmetric reinforcement design (same rebar in both end fiber of cross-section) is proposed because of existing of dynamic loads with changing direction (vibration, wind, earthquake, etc.) and unfavorable loading of live loads.

According to selected number of reinforcements, longitudinal rebar is assigned to candidate values and these values are rounded to fixed rebar sizes found in market. After the selection of longitudinal rebar, the fourth constraint related with minimum clear distance between bars is checked to ensure adherence. In g_4, ϕ is taken as the maximum size of candidate reinforcements and D_{max} is maximum size of aggregate used in concrete. Longitudinal rebar can be also positioned in two lines in end fiber.

Table 2. Design constraints

Description	Constraints
Maximum shear force (V_{nmax})	g_1: $V_d \leq V_{nmax}$=min{5.5bh; $0.2f'_c bh$ }
Maximum axial force (N_{max})	g_2: $N_d \leq N_{max}$= $0.5f'_c bh$
Maximum steel bars spacing, $a_{\phi max}$	g_3: $a_\phi \leq a_{\phi max}$=150 mm
Minimum steel bars spacing, $a_{\phi min}$	g_4: $a_\phi \geq a_{\phi min}$=max{1.5ϕ; 40 mm; (4/3)D_{max}}
Minimum steel area, A_{smin}	g_5: $A_s \geq A_{smin}$ = 0.01bh
Maximum steel area, A_{smax}	g_6: $A_s \leq A_{smax}$ = 0.06bh (seismic design)
Flexural strength capacity, M_{dx} and M_{dy}	g_7: $M_{dx} \geq M_{ux}$ and $M_{dy} \geq M_{uy}$
Concrete cover, c_c	g_8: $c_c \geq$ 30 mm
Axial force capacity, N_d	g_9: $N_d \geq N_u$
Shear strength capacity, V_d	g_{10}: $V_d \geq V_u$
Minimum shear reinforcement area, A_{vmin}	g_{11}: $A_v \geq A_{vmin}$=(bs/3f_y)
Maximum shear reinforcement spacing, s_{max}	g_{12}: s ≥ s_{max}=d/2 or d/4 if $V_s \geq 0.33\sqrt{f'_c}bd$

On assignment of rebar, the randomizations of bars in the direction with maximum flexural moment value is firstly done. Then, additional reinforcements are needed to randomly provide if the constraint; g_3 is not satisfied for the other direction. In addition to constraints ensuring the placement conditions, minimum and maximum reinforcement areas, respectively given as g_5 and g_6 must be also checked. Assignment of longitudinal reinforcements is iteratively conducted until constraints; g_3-g_6 are provided.

Then, flexural moment capacity is calculated, and it is compared with the required flexural moment for constraint g_7. In the analyses, the effective depth (d) of the column is taken by considering maximum size of randomly assigned longitudinal rebar and clear cover of the column defined as g_8. In the analysis stage, an iterative analysis is done to find distance from extreme compressive fiber to neutral axis (c) for axial forces and flexural moment in two directions. Then, deformations of the column are calculated for both directions and superposed. Finally, location of the neutral axis is found. After c is known, the exact axial force and flexural moment capacities are found and compared with the design loads (constraints; g_7 and g_9). If the moment and flexural moment conditions are not met, the optimum cost of the design is assigned with a penalized value like 10^6 money units. If these are provided, shear reinforcements are randomized and shear force capacity (g_{10}), minimum shear reinforcement area (g_{11}) and minimum shear reinforcement spacing (g_{12}) are checked. V_s is the shear force capacity provided by shear reinforcement in the constraint equations.

After all design variables are assigned with values providing all constraints defined according to ACI-318, the optimum material cost of the column defined as objective function (f(X)) for minimization is calculated as Eq. (4). In the objective function, u_{st} is length of shear reinforcement.

$$f(X) = \left(bh - \sum A_s\right) l C_c + \left(\sum A_s + \frac{A_v}{s} u_{st}\right) l \gamma_s C_s \tag{4}$$

After initial harmony memory matrix is generated, this matrix is modified according to rules of HS algorithm. The same methodology is assigned for these iterative cases. In global optimization, generation of design variables are done similar to Eq. (2). If possibility defined as harmony memory considering rate (HMCR) chosen between 0 and 1 at the start of optimization is smaller than a random number between 0 and 1, modification of HM matrix is done according to local search rules.

In optimum design of RC members, the selection of discrete design variables is proposed since dimensions can be provided with fixed sizes. The proposed value for dimensions is 50 mm differences and size of reinforcements are randomized with 2 mm differences for providing fixed market sizes. Also, the design constraints may not be provided if the randomizations are done around an existing vector. As the modification in the algorithm, the ranges of design variables are updated according to the best design and a classical PAR value is not used. The best value is assigned as minimum or maximum bound of the range. Both options may have 50% probability and this probability can be assumed as PAR. As all metaheuristic algorithms, the worst one is replaced with newly generated vector if cost of newly generated harmony vector is lower than existing one. The iteration of generating a new vector continues for maximum iteration number.

The Optimization Results

The optimization cases are presented as given in Table 3. In all cases, three different axial forces were investigated, and the values are 1000 kN, 1500 kN and 2000 kN. In all cases, the shear force value is taken as 100 kN. The numerical values of design constants and ranges of design variables are given in Table 4. The cost of materials changes according to the region of the construction. For that reason, the cost of steel for 1 ton is taken as 10 times of concrete per unit m^3. The optimum results for different axial forces are presented in Tables 5-7.

For 1000 kN axial force, additional reinforcements are not needed in Case 1. When the flexural moment values increase, web reinforcements are needed, and web reinforcements are generally assigned with minimum diameter sizes for cases 2-10. For the cases 11-12, the flexural moment in both directions are big and minimum web reinforcements is not enough.

For the optimum results given for 1500 kN axial force, a significant increase of optimum cross-sectional dimensions can be clearly seen. Since concrete is a cheap material respect to steel, additional compressive stresses resulting from axial forces are supported by concrete sections. In addition to that, increase of axial force and compressive stresses have a little advantage on reduction of tensile stresses, and the optimum costs are near to the costs of 1000 kN axial force cases for the cases 2, 7 and 12. For 2000 kN axial force cases, a significant reduction in the optimum total cost values is also seen for the cases 4-12 compared to the cases having 1500 kN axial force.

The optimum analysis results are done for 3 different axial force and different cases of biaxial bending moments are investigated. In the first 10 cases, a minor bending moment is used in one direction while the moment

Table 3. Biaxial flexural moments for different cases

Case	M_y (kNm)	M_x (kNm)
1	100	100
2	200	100
3	300	100
4	400	100
5	500	100
6	600	100
7	700	100
8	800	100
9	900	100
10	1000	100
11	500	500
12	800	800

Table 4. Design constants and ranges of design variables

Symbol	Unit	Value
b	mm	300-600
h	mm	300-600
l	mm	3000
ε_c	-	0.003
ϕ	mm	16-30
ϕ_v	mm	8-14
D_{max}	mm	16
f_y	MPa	420
f_c'	MPa	25
E_s	MPa	200000
γ_s	t/m^3	7.86
γ_c	t/m^3	2.5
C_c	unit/t	40
C_s	unit/ m^3	400

Table 5. Optimum design of biaxial loaded columns for 1000 kN axial force

Case No	b (mm)	h (mm)	Bars in Each Face (Critical Direction)	Web Reinforcement in Each Face (the Other Direction)	Shear Reinforcement Diameter/Distance (mm)	Total Cost
1	350	350	1Φ18+1Φ16+1Φ20	-	Φ8/150	32.57
2	300	500	1Φ22+1Φ18	2Φ16	Φ8/220	40.85
3	300	600	1Φ18+1Φ16+1Φ22	1Φ16	Φ8/270	44.32
4	300	600	2Φ24+ 1Φ16+1Φ20	1Φ16	Φ8/270	55.20
5	400	600	1Φ16+ 2Φ18+1Φ22+1Φ20	1Φ18+1Φ16	Φ8/270	67.23
6	350	600	5Φ18+2Φ20+1Φ30	1Φ16	Φ8/270	81.11
7	500	600	3Φ20+1Φ16+1Φ18+3Φ22	1Φ16	Φ8/240	91.54
8	550	550	4Φ22+1Φ26+1Φ28+1Φ16+1Φ18	1Φ16+1Φ18	Φ8/210	107.81
9	500	600	5Φ22+1Φ24+2Φ18+1Φ20+3Φ16	1Φ16+1Φ18	Φ8/240	119.39
10	500	600	1Φ20+3Φ24+1Φ22 2Φ28+1Φ18+1Φ30	3Φ18+1Φ16	Φ8/240	137.53
11	550	600	1Φ20+1Φ30+1Φ18+1Φ26	1Φ22+1Φ18	Φ8/210	90.15
12	600	600	1Φ30+ 1Φ18+1Φ16+4Φ26+1Φ20	3Φ22+1Φ30	Φ8/200	150.34

(Nigdeli et al., 2015)

Table 6. Optimum design of biaxial loaded columns for 1500 kN axial force

Case No	b (mm)	h (mm)	Bars in Each Face (Critical Direction)	Web Reinforcement in Each Face (the Other Direction)	Shear Reinforcement Diameter/Distance (mm)	Total Cost
1	450	300	3Φ18+1Φ16	-	Φ8/120	38.43
2	350	450	3Φ18+1Φ16	-	Φ8/190	40.66
3	400	500	3Φ18+ 1Φ16	1Φ16	Φ8/220	49.69
4	300	600	1Φ24+ 2Φ22	4Φ16	Φ8/270	62.58
5	500	550	1Φ22+ 1Φ24+1Φ30+1Φ20	1Φ18	Φ8/240	76.57
6	500	600	2Φ24+1Φ22+2Φ18	3Φ16	Φ8/240	85.16
7	550	600	2Φ16+3Φ22+1Φ20	2Φ16+1Φ18	Φ8/210	91.50
8	500	600	2Φ26+2Φ28+1Φ22+1Φ16+1Φ18	3Φ16	Φ8/240	110.03
9	550	600	10Φ18+6Φ16	2Φ16	Φ8/210	122.04
10	550	600	3Φ22+2Φ16+2Φ20 2Φ18+2Φ24+1Φ26+1Φ30	2Φ16	Φ8/210	141.99
11	600	600	1Φ20+ 1Φ22+1Φ26	1Φ24+1Φ28	Φ8/200	91.60
12	600	600	3Φ22+ 2Φ20+3Φ26+1Φ24+1Φ18	2Φ20+1Φ24+1Φ16	Φ8/200	148.53

(Nigdeli et al., 2015)

Table 7. Optimum design of biaxial loaded columns for 2000 kN axial force

Case No	b (mm)	h (mm)	Bars in Each Face (Critical Direction)	Web Reinforcement in Each Face (the Other Direction)	Shear Reinforcement Diameter/Distance (mm)	Total Cost
1	400	400	2Φ20	1Φ18	Φ8/170	39.69
2	350	500	1Φ22+1Φ16	2Φ16	Φ8/220	43.07
3	400	500	3Φ18+ 1Φ16	1Φ16	Φ8/220	49.69
4	350	600	1Φ22+ 1Φ28+1Φ16	1Φ16	Φ8/270	54.84
5	450	600	3Φ18+ 1Φ22+1Φ16	2Φ18	Φ8/260	70.90
6	350	600	3Φ20+2Φ24+1Φ22+ 1Φ18	1Φ18+1Φ16	Φ8/270	83.51
7	500	600	1Φ28+3Φ20+1Φ26+ 1Φ16	2Φ18	Φ8/240	92.65
8	500	600	8Φ18+1Φ16+1Φ20	1Φ16	Φ8/240	103.48
9	450	600	1Φ30+4Φ20+1Φ22+ 2Φ26+1Φ24	2Φ16	Φ8/260	119.47
10	550	600	2Φ26+1Φ28+1Φ24+ 2Φ22+1Φ20+2Φ18	3Φ16+1Φ18	Φ8/210	129.82
11	550	600	1Φ28+ 1Φ18+1Φ30+1Φ26	1Φ20	Φ8/210	89.80
12	600	600	1Φ20+ 5Φ18+2Φ28+2Φ24	2Φ28+ 1Φ26+1Φ18	Φ8/200	162.86

(Nigdeli et al., 2015)

in the other direction is increased between 100 kNm and 1000 kNm. As seen the results of these cases, the essential reinforcements are assigned in only critical direction and reinforcements with small diameters are placed for other direction in order to ensure the maximum clear distance between the reinforcement bars. Since the effect of biaxial bending is not clearly seen in these cases, additional two cases are also presented. In these cases, the flexural moment is equal for two directions and reinforcements are near to each other in both directions. Due to this situation, it is clearly seen that the increase of loads is not providing an optimum cost increase. This situation proves the importance of the optimization of RC columns.

OPTIMUM DESIGN OF UNIAXIAL LOADED COLUMNS

As a simple example of optimum RC design, an optimization code provided via Matlab (2018) is presented in Appendix 1. In this code, Jaya algorithm developed by Rao (2016) and design constraints is defined via ACI-318. In this study, detailed reinforcements are not given and only optimum reinforcement ratio (ρ) is optimized. The other design variables are breadth (b) and height (h) of RC section. The reinforcement ratio is taken as the same for both end fibers of section. In that case, ρ is equal to Eq. (5), where A_s and A_s' represent reinforcement area in end fibers.

$$\rho = \frac{A_s + A_s'}{bh} \tag{5}$$

In this example, design constraints; g_2, g_5, g_6, g_7 for one flexural moment and g_9 given in Table 2 are only checked.

The capacities of axial force (N_d) and flexural moment (M_d) are found according to Eqs. (6) and (7), respectively. A rectangular stress block with a distance is used as seen in Fig. 2. In this figure, ε_s, ε_s', ε_u, e, N, f_c', f_s, f_s', c, d, d′ and h represent strain of steel in compression section, strain of steel in tensile section, eccentricity of axial force, axial force, compressive strength of concrete, stress of steel in tension, stress of steel in compression, neutral axis distance, depth, distance from reinforcement center to end of section and height, respectively.

$$N_d = 0.85 f_c' ab + A_s' f_s' - A_s f_s \tag{6}$$

$$M_d = 0.85 f_c' ab \left(\frac{h}{2} - \frac{a}{2}\right) + A_s' f_s' \left(\frac{h}{2} - d'\right) + A_s f_s \left(d - \frac{h}{2}\right) \tag{7}$$

The flexural moment resulting from eccentricity can be formulated as follows:

$$M = Ne \tag{8}$$

The numerical values of design constants, design variable ranges and algorithm parameters are given in Table 8.

Detailed information about JA is found in the second chapter of this book. The cost of column is used as objective function is similar to Eq. (4), but shear reinforcements are not considered as seen Eq. (9).

$$f(X) = \left(bh - \sum A_s\right) l C_c + \left(\sum A_s\right) l \gamma_s C_s \tag{9}$$

The optimum results are presented as Table 9 and continuous design variables are used. In the optimum results, ρ was minimized and h was maximized while optimum b value was found.

OPTIMIZATION OF RC COLUMNS BY CONSIDERING SLENDERNESS

In several design codes and ACI 318, second order effects of RC columns are considered by using an approximate design procedure magnifying flexural moments with moment magnification factor (δ_s). In this section, the formulation of calculation of moment magnification factor is given. In the calculation, moment of inertia of members is reduced by 65% and 30% for beam and columns, respectively. In that case, the cracking of members is considered. The moment magnification factor is calculated according to slenderness of columns. In that case, effective length factors in buckling (*k*) are calculated for all columns. In calculations, Ψ at both ends of columns (Ψ_A at upper

Table 8. Design constant and ranges of design variables

Definition	Value
Range of web width, bw	250 mm-400 mm
Range of height, h	300 mm-600mm
Ratio of reinforcement (ρ)	0.01-0.06
Yield strength of steel, f_y	420 MPa
Comp. strength of concrete, f'_c	25 MPa
Elasticity modulus of steel, E_s	200000 MPa
Specific gravity of steel, γ_s	7.86 t/m³
Cost of the concrete per m³	40 unit
Cost of the steel per ton	400 unit
distance from reinforcement center to end of section, d'	60 mm
Maximum strain at extreme concrete compression fiber, ε_{cu}	0.003
Population number, pn	20
Maximum iteration number, maxiter	20000
Length of the column, *l*	3m
Axial force, N	2000 kN
Flexural moment, M	50 kNm

Table 9. Optimum results with JA

Width of column (b)	279.3711 mm
Height of column (h)	600 mm
Reinforcement ratio (ρ)	0.01
Optimum cost (unit)	35.92488

and Ψ_B at lower end) given in Eq (10) is used. E, I, and *l* represent elasticity modulus, moment of inertia and length of the RC members, respectively.

$$\Psi_{A,B} = \frac{\sum \left(EI/l \right)_{column}}{\sum \left(EI/l \right)_{beam}} \tag{10}$$

k is calculated according to lateral displacement ability of the structural system. For structures which are free to make lateral displacements, k is calculated as follows:

Figure 2. Column under uniaxial loading

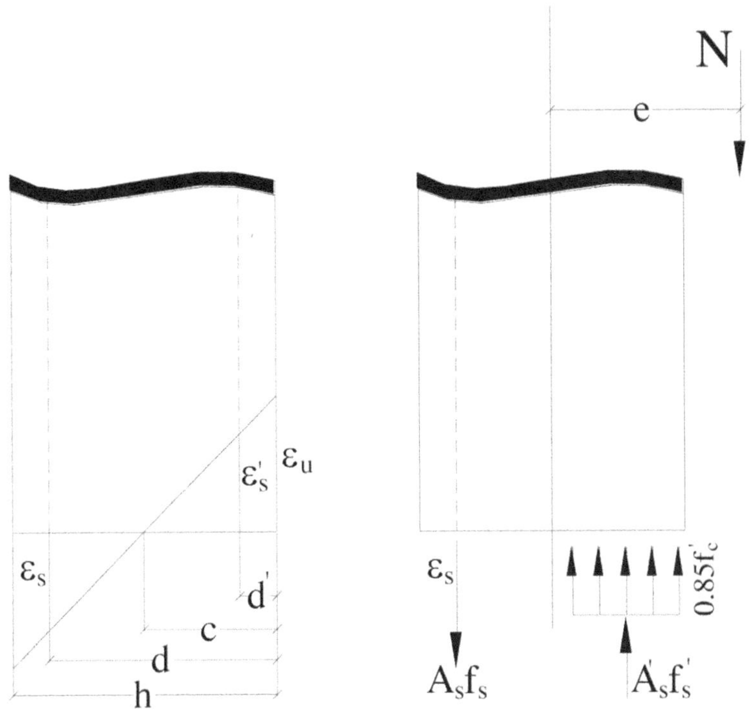

$$\Psi_m = 0.5(\Psi_A + \Psi_B) \tag{11}$$

$$k = \frac{20 - \Psi_m}{20}\sqrt{1 + \Psi_m}\ if\ \Psi_m < 2 \tag{12}$$

$$k = 0.9\sqrt{1 + \Psi_m}\ if\ \Psi_m \geq 2 \tag{13}$$

If the lateral displacement is restricted, k is calculated as follows if Ψ_{min} is the minimum of Ψ_A and Ψ_B.

$$k = \min\begin{Bmatrix} 0.7 + 0.05(\Psi_A + \Psi_B) \leq 1 \\ 0.85 + 0.05\Psi_{\min} \leq 1 \end{Bmatrix} \tag{14}$$

The equation of moment magnification factor (δ_s) is shown in Eq (15).

$$\delta_s = \frac{C_m}{1 - \frac{P_u}{0.75P_c}} \tag{15}$$

C_m represents the correction factor used for considering actual moment diagram to end equivalent moment diagram. It is calculated as given in Eq (16). C_m must be taken as bigger than 0.4. For member with transverse loads between supports, it is taken as 1.0.

$$C_m = 0.6 + 0.4\frac{M_1}{M_2} \tag{16}$$

Critical buckling load is calculated as follows and the rigidity of column is reduced by 75% in the calculation:

$$P_c = \frac{\pi^2 EI}{(kl)^2} \tag{17}$$

The optimum design of RC columns considering slenderness was done in several studies. In these studies, Harmony Search (HS) algorithm (Bekdaş & Nigdeli, 2014), Bat Algorithm (BA) (Bekdaş & Nigdeli, 2016a) and Teaching Learning Based Optimization (TLBO) (Bekdaş & Nigdeli, 2016b) were employed. The optimum results are presented in Table 11 for design constant and design variable ranges given as Table 10. The optimum results are given for different lengths and the external loads taken in the optimization are 2000 kN, 50 kNm and 50 kN for axial force, flexural moment and shear force, respectively.

The slenderness effect can be seen from the optimum results. Although for all length cases, the external forces are the same, the increase of the total cost does not show a linear increase for long columns.

Table 10. Design constant and ranges of design variables

Definition	Value
Range of web width, b_w	250 mm-400 mm
Range of height, h	300 mm-600mm
Longitudinal reinforcement (ϕ)	16 mm-30 mm
Shear reinforcement (ϕ_v)	8 mm-14 mm
effective length factor in buckling, k	1.2
Clear cover, c_c	30mm
Max. aggregate diameter, D_{max}	16 mm
Yield strength of steel, f_y	420 MPa
Comp. strength of concrete, f'_c	25 MPa
Elasticity modulus of steel, E_s	200000 MPa
Specific gravity of steel, γ_s	7.86 t/m^3
Specific gravity of concrete (γ_c)	2.5 t/m^3
Cost of the concrete per m^3	40 unit
Cost of the steel per ton	400 unit

Table 11. Design constant and range

	HS			BA			TLBO		
Length of the column (l)	3m	4m	5m	3m	4m	5m	3m	4m	5m
Breadth of column (b_w) (mm)	400	300	300	400	300	300	400	300	300
Height of column (h) (mm)	400	550	600	400	550	600	400	550	600
Bars in each face	1Φ20+1Φ18	2Φ16	2Φ16	3Φ16	2Φ16	2Φ16	3Φ16	2Φ16	2Φ16
Web reinforcement in each face	1Φ18	1Φ16+1Φ18	2Φ18	1Φ16	1Φ16+1Φ18	2Φ18	1Φ16	1Φ16+1Φ18	2Φ18
Shear reinforcement diameter (mm)	Φ8	Φ8	Φ8	Φ8	Φ8	Φ8	Φ8	Φ8	Φ8
Shear reinforcement distance (mm)	170	240	270	170	240	270	170	240	270
Optimum cost (unit)	38.58	52.27	69.97	38.22	52.27	69.97	38.22	52.27	69.97

REFERENCES

ACI-318. (2005). Building code requirements for structural concrete and commentary, metric version, *American Concrete Institute*.

Bekdas, G., & Nigdeli, S. M. (2014). The optimization of slender reinforced concrete columns. *Proceedings in Applied Mathematics and Mechanics*, *14*(1), 183–184. doi:10.1002/pamm.201410079

Bekdas, G., & Nigdeli, S. M. (2016a). Bat algorithm for optimization of reinforced concrete columns. *Proceedings in Applied Mathematics and Mechanics*, *16*(1), 681–682. doi:10.1002/pamm.201610329

Bekdaş, G., & Nigdeli, S. M. (2016b, September). Optimum design of reinforced concrete columns employing teaching learning based optimization. In *Proceedings 12th International Congress on Advances in Civil Engineering*, (pp. 1-8). Academic Press.

Geem, Z. W., Kim, J. H., & Loganathan, G. V. (2001). A new heuristic optimization algorithm: Harmony search. *Simulation*, *76*(2), 60–68. doi:10.1177/003754970107600201

Mathworks. (2018). MATLAB R2018a. The MathWorks Inc., Natick, MA.

Nigdeli, S. M., Bekdas, G., Kim, S., & Geem, Z. W. (2015). A novel harmony search-based optimization of reinforced concrete biaxial loaded columns. *Structural Engineering and Mechanics*, *54*(6), 1097–1109. doi:10.12989em.2015.54.6.1097

Rao, R. (2016). Jaya: A simple and new optimization algorithm for solving constrained and unconstrained optimization problems. *International Journal of Industrial Engineering Computations*, *7*(1), 19–34.

APPENDIX

```
%DESIGN OPTIMIZATION OF REINFIRCED CONCRETE RECTANGULAR COLUMN
WITH JAYA ALGORITHM
clear all
clc
%1.ENTER DATA OF PROBLEM
fc_prime=25;          % specified compressive strength of
concrete,[Mpa]
fy=420;               % specified yield strength of
reinforcement,[Mpa]
gama_steel=7.86;      % unit volume weight of stell, [t/m^3]
Es=200000;            % modulus of elasticity of reinforcement,
[Mpa]
eps_cu=0.003;         % Maximum usable strain at extreme
concrete compression fiber
L=3000;               % length of column, [mm]
d_prime=60;           % distance from extreme compression fiber
to centroid of longitudinal compression reinforcement,[mm]
Mu=50*10^6;           % factored moment at section, [N*mm]
Nu=2000*10^3;         % factored axial force normal to cross
section, [N]
Cc=40;                % unit cost of concrete [unit/m^3]
Cs=400;               % unit cost of reinforcement [unit/t]
% factor relating depth of equivalent rectangular compressive
stress block to neutral axis depth, ACI318 (Section 10.2.7.3)
if fc_prime<=28
   beta1=0.85;
end
if fc_prime>28
   beta1=0.85-0.05*(fc_prime-28)/7;
end
if beta1<0.65
   beta1=0.65;
end
% User defined algorithm parameters
pn=20;          % population number
maxiter=2000; % maximum iteration number
% Ultimate limits of design variables
% bmin: lower limit for width of cross-section
% bmax: upper limit for width of cross-section
% hmin: lower limit for height of cross-section
% hmax: upper limit for height of cross-section
% romin: lower limit for ratio of reinforcement
% romax: upper limit for ratio of reinforcement
bmin=250  ; bmax=400;
hmin=300  ; hmax=600;
```

```
romin=0.01; romax=0.06;
%2. GENERATING OF INITIAL SOLUTION MATRIX
% b  : width of cross-section [mm]
% h  : height of cross-section [mm]
% ro : ratio of reinforcement
% rand: a function generating uniformly distributed
pseudorandom numbers in MATLAB
for i=1:pn % During the this loop, candidate solution vectors
including design variables are generated up to pn.
          % End of the this loop, candidate solution vectors
providing design constraint are stored in a matrix.
    % 2.1. DEFINITION OF DESING VARIABLES RANDOMLY
    b=bmin+rand*(bmax-bmin);      % Generation of b between
ultimate limits and randomly
    h=hmin+rand*(hmax-hmin);      % Generation of h between
ultimate limits and randomly
    ro=romin+rand*(romax-romin); % Generation of reinforcement
ratio between ultimate limits and randomly
    % 2.2. OTHER CALCULATIONS
    As=ro*b*h/2;        % Area of nonprestressed longitudinal
tension reinforcement, [mm^2]
    As_prime=ro*b*h/2; % Area of longitudinal compression
reinforcement, [mm^2]
    % Location of columns and total area of reinforments are
stored in matrix K:
    K(1,1)=As;
    K(1,2)=d_prime;
    K(2,1)=As_prime;
    K(2,2)=h-d_prime; % Distance from extreme compression fiber
to centroid of longitudinal tension reinforcement,[mm]
    % Starting and ending points of the neutral axis are
determined:
    cx_bas=min(K(:,2)); % Selection of minimum value in the
second column of matrix K
    cx_bit=max(K(:,2)); % Selection of maximum value in the
second column of matrix K
    % NOTES:
    % These ultimate limit points can be select different from
these values.
    % min: command finding the mimimum value in any vector
    % max: command finding the maximum value in any vector
    % Objective function assigned as an penaltized value (10^6)
before calculation of column capacity:
    F_obj=10^6;
    % Calculation of column capacity or generating moment-axial
force interaction diagram:
    for c=cx_bas:cx_bit
        a=c*beta1;      % depth of equivalent rectangular stress
block as defined in ACI318-(Section 10.2.7.1) [mm]
```

```
        comp_area=b*a; % total cocrete area under compression
strength effect according to equivalent rectangular stress
block theory
        CM=h/2-a/2;    % distance ftrom compression area to
center of gravity
        for j=1:numel(K(:,2))
            % reinforcements are multiplied by the defromations
of the reinforcement:
            % NOTE: In the calculations, if reinforcement
reaches the yield limit fy is used, otherwise eps_cu*(c-d_
prime/c) is used.
            if (c-K(j,2))>0
                K(j,3)=K(j,1)*min(fy,Es*(eps_cu*(c-K(j,2))/c));
% compressive forces
            else
                K(j,3)=-K(j,1)*min(fy,Es*abs(eps_cu*(c-
K(j,2))/c)); % tensile forces
            end
            %Calculation of bending moments carried by
reinforcements:
            K(j,4)=(h/2-K(j,2))*K(j,3);
        end
        Nn=(0.85*fc_prime*comp_area+sum(K(:,3)));    % nominal
strength in tension, [N]
        Mn=(0.85*fc_prime*comp_area*CM+sum(K(:,4))); % nominal
flexural strength at section, [Nmm]
        % sum: command finding the total value in any vector
        % Calculation of strength reduction factor (phi)
according to ACI318- Sections 9.3.2.2 and 10.3.3
        for j=1:numel(K(:,2))
            if (c-K(j,2))<0
               if abs(eps_cu*(c-K(j,2))/c)<= 0.002
                  phi=0.65;
               end
               if 0.002 < abs(eps_cu*(c-K(j,2)))/c && abs(eps_
cu*(c-K(j,2)))/c< 0.005
                    phi=0.65+(0.25/0.003)*(abs(eps_cu*(c-
K(j,2))/c)-0.002);
               end
               if 0.005 <= abs(eps_cu*(c-K(j,2)))/c
                  phi=0.90 ;
               end
            end
        end
        % 2.3. OBJECTIVE FUNCTION
        % NOTE: Objective function (total cost) was assigned as
an penaltized value (10^6) before calculation of column
capacity.
        % If column capacity provides external loads objective
```

```
function recalcultaed according to concerete and steel amounts.
        % In other words, the objective function is
recalculated if the Mn and Nn forces remain in the diagram:
        if phi*Nn<=0.5*fc_prime*b*h && phi*Nn>=Nu && phi*Mn>=Mu
           F_obj=L*b*h*Cc/(10^9)+gama_steel*L*(As+As_prime)*Cs/
(10^9);
        end
    end
    % 2.4. INITIAL SOLUTION MATRIX
    % NOTE: End of the each loop, a candidate solutution vector
including randomly generated design variables stored in the
matrix named OPT
    % Thus each vector in matrix is candidate solution vector.
    OPT(1,i)=b;
    OPT(2,i)=h;
    OPT(3,i)=ro;
    OPT(4,i)=F_obj;
end
% 3. GENERATING OF NEW SOLUTION MATRIX VIA JA
% NOTE: This part is very similar to PART 2. Differently, new
solution variables generated acoording to JA algorithm and
ultimate limits of variables
% are controlled. Additionaly, objective function are
recalculated in case of sufficiant column capacity under
external loads and
% reinforcement ratio that is less than ultimate limit.
% All steps in the PART3 are repeated until all iteration is
completed.
for iter=1:maxiter
    for i=1:pn
        % 3.1. DEFINATION OF DESING VARIABLES
        % Values of candidate solution vector provided best/
minimum value of objective function in existing solution set or
matrix are assigned to vector named best:
        [pe,re]=min(OPT(4,:));
        best(1,1)=OPT(1,re);
        best(2,1)=OPT(2,re);
        best(3,1)=OPT(3,re);
        % Values of candidate solution vector providing worst/
maximum value of objective function in existing solution set or
matrix are assigned to vector named worst:
        [ke,te]=max(OPT(4,:));
        worst(1,1)=OPT(1,te);
        worst(2,1)=OPT(2,te);
        worst(3,1)=OPT(3,te);
        % New values of design variables are generated
according to JA algorithm
        b=OPT(1,i)+rand*(best(1,1)-abs(OPT(1,i)))-
rand*(worst(1,1)-abs(OPT(1,i)));
```

```
        h=OPT(2,i)+rand*(best(2,1)-abs(OPT(2,i)))-
rand*(worst(2,1)-abs(OPT(2,i)));
        ro=OPT(3,i)+rand*(best(3,1)-abs(OPT(3,i)))-
rand*(worst(3,1)-abs(OPT(3,i)));
        % 3.2. CONTROL OF ULTIMATE LIMITS
        if b>bmax
           b=bmax;
        end
        if b<bmin
           b=bmin;
        end
        if h>hmax
           h=hmax;
        end
        if h<hmin
           h=hmin;
        end
        if ro<romin
           ro=romin;
        end
        % 3.3. OTHER CALCULATIONS
         As=ro*b*h/2;       % Area of nonprestressed
longitudinal tension reinforcement, [mm^2]
         As_prime=ro*b*h/2; % Area of longitudinal compression
reinforcement, [mm^2]
        % Location of columns and total area of reinforments
are stored in matrix K:
        K(1,1)=As;
        K(1,2)=d_prime;
        K(2,1)=As_prime;
        K(2,2)=h-d_prime; % Distance from extreme compression
fiber to centroid of longitudinal tension reinforcement,[mm]
        % Starting and ending points of the neutral axis are
determined:
        cx_bas=min(K(:,2));
        cx_bit=max(K(:,2));
        % NOTE: These points can be select different from these
values.
        % min: command finding the mimimum value in any vector
        % max: command finding the maximum value in any vector
        % Objective function assigned as an panlized value
(10^6) before calculation of column capacity:
        F_obj=10^6;
       % Calculation of column capacity or generating moment-
axial force interaction diagram:
       for c=cx_bas:cx_bit
           a=c*beta1;      % depth of equivalent rectangular
stress block as defined in ACI318 (Section 10.2.7.1) [mm]
           comp_area=b*a; % total cocrete area under
```

```
compression strength effect according to equivalent rectangular
stress block theory
          CM=h/2-a/2;    % distance ftrom compression area to
center of gravity
          for j=1:numel(K(:,2))
              % According to the law of hook, the total cross-
sectional area and the total force carried by the
              % reinforcements are multiplied by the
defromations of the reinforcement:
              % NOTE: In the calculations, if reinforcement
reaches the yield limit fy is used, otherwise eps_cu*(c-d_
prime/c) is used.
              if (c-K(j,2))>0
                  K(j,3)=K(j,1)*min(fy,Es*(eps_cu*(c-
K(j,2))/c));     % compressive forces
              else
                  K(j,3)=-K(j,1)*min(fy,Es*abs(eps_cu*(c-
K(j,2))/c)); % tensile forces
              end
              %Calculation of bending moments carried by
reinforcements:
              K(j,4)=(h/2-K(j,2))*K(j,3);
          end
          Nn=(0.85*fc_prime*comp_area+sum(K(:,3)));     %
nominal strength in tension, [N]
          Mn=(0.85*fc_prime*comp_area*CM+sum(K(:,4))); %
nominal flexural strength at section, [Nmm]
          % sum: command finding the total value in any vector
          % Calculation of strength reduction factor (phi)
according to ACI318 Sections 9.3.2.2 and 10.3.3
          for j=1:numel(K(:,2))
              if (c-K(j,2))<0
                 if abs(eps_cu*(c-K(j,2))/c)<= 0.002
                    phi=0.65;
                 end
                 if 0.002 < abs(eps_cu*(c-K(j,2)))/c &&
abs(eps_cu*(c-K(j,2)))/c< 0.005
                    phi=0.65+(0.25/0.003)*(abs(eps_cu*(c-
K(j,2))/c)-0.002);
                 end
                 if 0.005 <= abs(eps_cu*(c-K(j,2)))/c
                    phi=0.90 ;
                 end
              end
          end
          % 3.4. OBJECTIVE FUNCTION
          % NOTE: Objective function (total cost) was assigned
as an penaltized value (10^6) before calculation of column
capacity.
```

```
            % If column capacity provides external loads and
ultimate limit of reinforcement ratio,
            % objective function recalculated according to
concrete and steel amounts:
            if phi*Nn<=0.5*fc_prime*b*h && phi*Nn>=Nu &&
phi*Mn>=Mu
               F_obj=L*b*h*Cc/(10^9)+gama_steel*L*(As+As_
prime)*Cs/(10^9);
            end
        end
        % 3.5. NEW SOLUTION MATRIX
        % NOTE: End of the each loop, a new candidate
solutution vector including randomly generated design variables
stored in the matrix named OPT1
        % Thus each vector in matrix is candidate solution
vector.
        OPT1(1,i)=b;
        OPT1(2,i)=h;
        OPT1(3,i)=ro;
        OPT1(4,i)=F_obj;
        % 3.6. PART OF COMPARISON
        % NOTE: Each candidate solution vector in the new
solution matrix named OPT1 is compared with existing solution
vectors in maxtrix OPT
        % If new solution vectors provide better solution or
minimum cost replaced with existing solution vectors:
        if OPT(4,i)>OPT1(4,i)  % Comparing of values of
objective functions
            OPT(:,i)=OPT1(:,i); % Updating of matrix OPT
        end
    end
end
```

Chapter 5
Metaheuristic Optimization of Reinforced Concrete Footings:
Optimization of RC Footings

ABSTRACT

Footings are one of the structural members, which is one of the complex engineering problems to optimize. Differently from the other reinforced concrete member designs such as beams and columns, geotechnical limit states are also needed to consider in addition to structural state limits. In this chapter, the optimum design of RC footing is presented according to ACI 318: Building code requirements for structural concrete. The optimum results of methodologies employing different algorithms including Harmony Search (HS), Teaching-Learning-Based Optimization algorithm (TLBO), and Flower Pollination Algorithm (FPA). In the methodologies, the design variables are values about dimensions of the footing, the orientation of the supported columns and reinforcements. Also, a simplified methodology is also presented with a design code employing HS.

INTRODUCTION

As reinforced concrete (RC) columns, footings are also a major component of the structures since failure of them provides the total collapse of structure. Spread footings are the most basic type of footings and these footings are mounted at the bottom of basement columns, but in design of RC spread

DOI: 10.4018/978-1-7998-2664-4.ch005

footings, several structural design limits including the punching effect and geotechnical state limits related to capacity and properties of supporting soil. A safe design providing all strength against all types of stresses is not enough if the stability of footings is not provided according to soil bearing capacity. Comparing to other optimization cases of members, in additional to existing non-linear behavior of RC members, the footing design is more complicated than others since the design variables are both effective on geotechnical and structural state limits.

Nigdeli, Bekdaş and Yang (2018) optimized spread footings by employing several methodologies and the results and these calculations are formulated in this chapter. Differently from the most basic form of spread footing taken as rectangular, a trapezoidal shape to save from the volume of the concrete is presented by using biaxial flexural moments of the mounted column. As additional design variables, orientation (eccentricity respect to the midpoint of footing) of the column is considered to reduce internal forces and to provide minimization of objective function taken as the cost of design. Differently from dimension design variables, quantity of reinforcements in all critical sections are considered as design variables. All design constraints are defined according to ACI 318: Building code requirements for structural concrete. The algorithms with the presented optimum results are Harmony Search (HS) algorithm developed by Geem, Kim and Loganathan (2001), Teaching-Learning Based Optimization (TLBO) developed by Rao, Savsani, and Vakharia (2011), Flower Pollination Algorithm (FPA) developed by Yang (2012), Particle Swarm Optimization (PSO) developed by Kennedy and Eberhart (1995) and Differential Evolution (DE) developed by Storn and Price (1997). Then, a computer code is presented for a simplified methodology.

OPTIMUM DESIGN OF RC SPREAD FOOTINGS FOR COLUMNS UNDER BIAXIAL FLEXURAL MOMENT

The design of RC footings involves two different limit states, which are formulated as design constraints in the optimization process. These limit states are namely, geotechnical and structural limit states. These states are separately considered in order to save from the computational effort. Physical mechanic formulations and experimental results are formulated in design codes to represent these limit states.

In the presented methodology, the optimization process related to the assignment of candidate solutions for optimum reinforcement design is not conducted to save from computation time, when selected dimensions of footing is not suitable to provide geotechnical limit states.

As explained in previous chapters, an iterative random search process employing heuristics is done in two phases of optimization, namely global and local optimizations. By these two phases, trapping to a local optimum is prevented via a global search process. In addition to that, the convergence ability is provided via a local search process.

The design variables of spread footing are dimensions including base dimensions (x_1 and x_2), height of the footing (x_3) and orientation of the column defined as an eccentricity respect to midpoint of base of footing (x_4 and x_5), and detailed reinforcement design (x_6, x_7, x_8 and x_9). All design variables are shown in Fig. 1. In the drawings, C_f and C_c represents centers of footing and column, respectively. The base dimensions (x_1, x_2) are also denoted as L and B in x and y directions, respectively. The height of the footing taken at the face of the column (H) is the third-dimension design variable (x_3). The minimum height taken as the end-sides of footing is represented with a constant minimum value, H_{min}. In order to find an optimum balance of stress under footing resulting from biaxial flexural moments in x (M_x) and y (M_y) directions and axial force in additional to the axial force (P), dimensions related to eccentricity of the column respect to midpoint of footing are two distances; e_x and e_y which are the fourth (x_4) and fifth (x_5) design variables, respectively. The sizing parameters (diameters of bars in mm) of rebars are x_6 and x_7 in two sections defined as A-A and B-B. Lastly, distances between rebars are shown as x_8 and x_9. All design variables are suggested to be discrete variables because of ability of construction in construction yard. Especially, footings are elements with big dimensions and constructed in soil. For that reasons, rounded fixed dimensions and fixed rebar sizes found in local markets are essentially needed.

As all presented optimization methodologies, the design constants and ranges of design variable are first defined. Then, an initial matrix containing candidate values is generated. In first assigning of design variables, a random solution can be chosen within the range. Then, design variables are updated according to the formulation of algorithms presented in chapter 2 in each iteration by selecting a global or local optimization phase via a probability parameter. TLBO has no probability parameter and both phases are consequently applied.

Figure 1. The optimization problem with design variables
(Nigdeli et al., 2018)

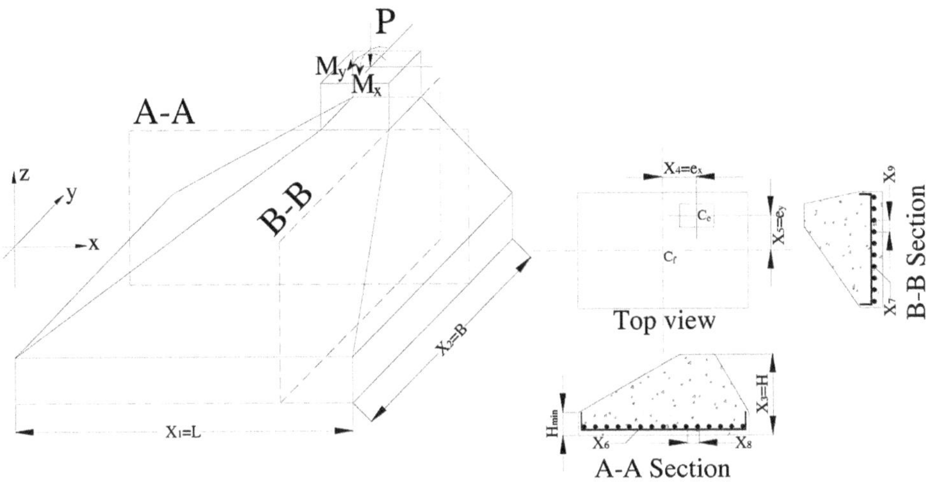

As written before, a multi-step procedure is proposed in the random generation process of design variables to save computational time. Firstly, the variables about dimensions (x_1-x_5) are defined. Then, the geotechnical limit states about bearing pressure on the soil and the settlement (δ) are checked, and the objective function (f_{cost}) is penalized if the related design constraints are violated.

The bearing pressures ($q_{1,2,3,4}$; from 1 to 4) are calculated at four sites of the foundation and the equation is formulated as Eq. (1). W_f is the total weight of the foundation (self-weight) including the soil on the top of the foundation.

$$q_{1,2,3,4} = \frac{P + W_f}{BL} \pm \frac{6(M_y - Pe_x)}{BL^2} \pm \frac{6(M_x - Pe_y)}{B^2 L} \tag{1}$$

By using these extremum bearing pressures, two state limit check is done. For each one, the stability of the foundation is checked by providing the inequality constraint presented as Eq. (2). Since soil cannot carry tensile forces, all pressures must be over 0. Secondly, a factor of safety (FS) is defined for the maximum pressure as given in Eq. (3).

$$q_{1,2,3,4} \geq 0 \tag{2}$$

$$FS < \frac{q_{ult}}{\max(q_{1,2,3,4})} \tag{3}$$

For a cohesionless soil, ultimate bearing capacity of soil (q_{ult}) with no ground slope is calculated as Eq. (4).

$$q_{ult} = \gamma D N_q F_{qs} F_{qd} + 0.5\gamma B N_\gamma F_{\gamma s} F_{\gamma qd} \tag{4}$$

In the bearing capacity equation, γ is the unit weight of the soil. Eq. (4) includes several factors such as bearing capacity factors (N_q and N_γ) and shape depth factors (F_{qs}, $F_{\gamma s}$, F_{qd}, $F_{\gamma d}$). These factors are formulated as Eqs. (5-10). The internal friction angle is shown with ϕ'.

$$N_q = e^{\pi \tan\varphi'} \tan^2\left(\frac{\pi}{4} + \frac{\varphi'}{2}\right) \tag{5}$$

$$N_\gamma = 2(N_q + 1)\tan\varphi' \tag{6}$$

$$F_{qs} = 1 + \frac{B}{L}\tan\varphi' \tag{7}$$

$$F_{\gamma s} = 1 - 0.4\frac{B}{L} \tag{8}$$

$$F_{qd} = \begin{cases} 1 + 2\tan\varphi'(1-\sin\varphi')^2 \left[\arctan\left(\frac{D}{B}\right)\right] & if\ D > B \\ 1 + 2\tan\varphi'(1-\sin\varphi')^2 \left(\frac{D}{B}\right) & if\ D < B \end{cases} \tag{9}$$

$$F_{\gamma d} = 1.0 \tag{10}$$

Then, settlement (δ) of footing is calculated according to the elastic solution of Poulus and Davis (1974) as shown in Eq. (11) and value of settlement must be equal or lower than the maximum allowed settlement (δ_{max}) as a geotechnical state limit constraint. In Eq. (11), elasticity modulus and Poisson's ratio of soil are defined as E and ν, respectively. Also, shape factor denoted by β_z is calculated according to Whitman and Richart (1967) and the formulation is given as Eq. (12).

$$\delta = \frac{P + W_f(1-\nu^2)}{\beta_z E\sqrt{BL}} \leq \delta_{\max} \tag{11}$$

$$\beta_z = -0.0017\left(\frac{L}{B}\right)^2 + 0.0597\left(\frac{L}{B}\right) + 0.9843 \tag{12}$$

If one of the constraints about geotechnical state limits given as Eqs (2), (3) and (11) are provided, total cost taken as objective function is penalized and the further processes in an iteration are not applied. The penalized total cost is assigned with a big value like 10^6 money units used in finding the presented optimum results.

After the satisfaction of geotechnical limit states, the process continues by generating design variables about rebars. Then, the controlling structural state limits start. This process is generally like constraints of RC beams. Tension controlled based design (ductility condition allowing yielding of steel in tension before failure of concrete in compression) is considered for the maximum reinforcement area in flexure and the compressive block of the concrete can be taken as an equivalent rectangular stress block as done for obtaining the presented optimum results. After the calculation of flexural moment and shear force at critical sections, capacity of footing respect to these internal forces is checked. The critical sections are the points of footings, where column is mounted (Sides of column for flexure and a distance far as average effective depth; d_{ave} of footing for shear).

Differently from flexural moment capacity of two directions and shear force capacity, two-way shear capacity (punching) of the footing must be also checked, because the column tends to punch through the spread footing. The critical punching perimeter is taken from $d_{ave}/2$ away from the column face. It is formulated as Eq. 13 and b_{column} is the breadth of the column for corresponding direction.

$$b_{perim} = 4(b_{column} + d_{ave}) \tag{13}$$

The capacity of the RC footing for one-way shear ($V_{n,one\text{-}way}$) is given as Eq. (14). Additional stirrups are not used in a classical spread footing design. In that case, the capacity must provide the critical shear force. The width of footing is shown as *w* and it is taken as B or L according to directions x and y, respectively.

$$V_{n,one-way} = 0.75\left(0.17wd_{ave}\sqrt{f_c^{'}}\right) \tag{14}$$

Two-way shear ($V_{n,two\text{-}way}$) is calculated according to Eqs. (15). β represents the ratio of the long side to short side of the column. The compressive strength of concrete is given as f_c'.

$$V_{n,two-way} = 0.75 * \min\begin{cases} 0.17\left(1+\dfrac{2}{\beta}\right)\sqrt{f_c^{'}}b_{perim}d_{ave} \\ 0.083\left(\dfrac{4d_{ave}}{b_{perim}}+2\right)\sqrt{f_c^{'}}b_{perim}d_{ave} \\ 0.33\sqrt{f_c^{'}}b_{perim}d_{ave} \end{cases} \tag{15}$$

The objective function is penalized if one of the structural limit states (flexural moment, shear and punching) are not provided. If all geotechnical and structural limit states are provided, the objective function (f_{cost}) defined as the total cost of the footing is calculated as Eq. (16), where volume of concrete ($V_{concrete}$), cost of concrete for unit volume ($C_{concrete}$), the weight of the reinforcements (W_{steel}) and the cost of reinforcement for unit weight (C_{steel}) are used. The volume of concrete and the weight of steel is calculated according to values of design variables.

$$f_{cost} = V_{concrete} C_{concrete} + W_{steel} C_{steel} \quad (16)$$

A general flowchart of the methodology is given as Fig. 2.

The Numerical Cases and Optimization Results

The design constants and design variable ranges are given in Table 1. The dead and live loads of the cases are given in Table 2, and the axial force and flexural moments are factored with a ϕ value taken as 1.2 and 1.6 for dead and live loads, respectively. The dead and live loads are represented by adding subscripts G and Q, respectively. For applicable production in the construction yard, discrete design variables about dimensions are taken with 5cm differences.

Optimization was done for one million analyses, but all algorithms such HS, TLBO, FPA, PSO and DE are effective to find the same optimum result before maximum iteration number for Cases 1-4 as given in Table 3.

Cases 2 and 3 are similar cases and only the direction of flexural moments are flipped. As seen in the optimum results given as Table 3, the same results for cases 2 and 3 are obtained. This situation validates the robustness of the method.

To adjust the minimum and maximum pressure on the sides, a non-symmetric design was founded as optimum for Cases 1 and 4. The number of analyses for finding the optimum values are very different according to algorithms, and DE and FPA are the fastest ones.

For 50 independent runs of the methodology, standard deviation values are given, and t-test analyses are presented by taking PSO approach results as the reference. In Table 4, the t_{stat} values calculated according to Eq. (17) are shown for α=5% significance level and 49 degrees of freedom. In that case, t critical value (t_{cr}) is -1.6765. $\overline{x}$ represents the average value of the reference method (PSO), s is standard deviation and n is number of observations. According to the results given in Table 4, the average values (μ) of the methods are generally better than the PSO approach with 95% confidence level.

$$t_{stat} = \frac{\overline{x} - \mu}{s / \sqrt{n}} \quad (17)$$

Figure 2. The flowchart of the optimization methodology
(Nigdeli et al., 2018)

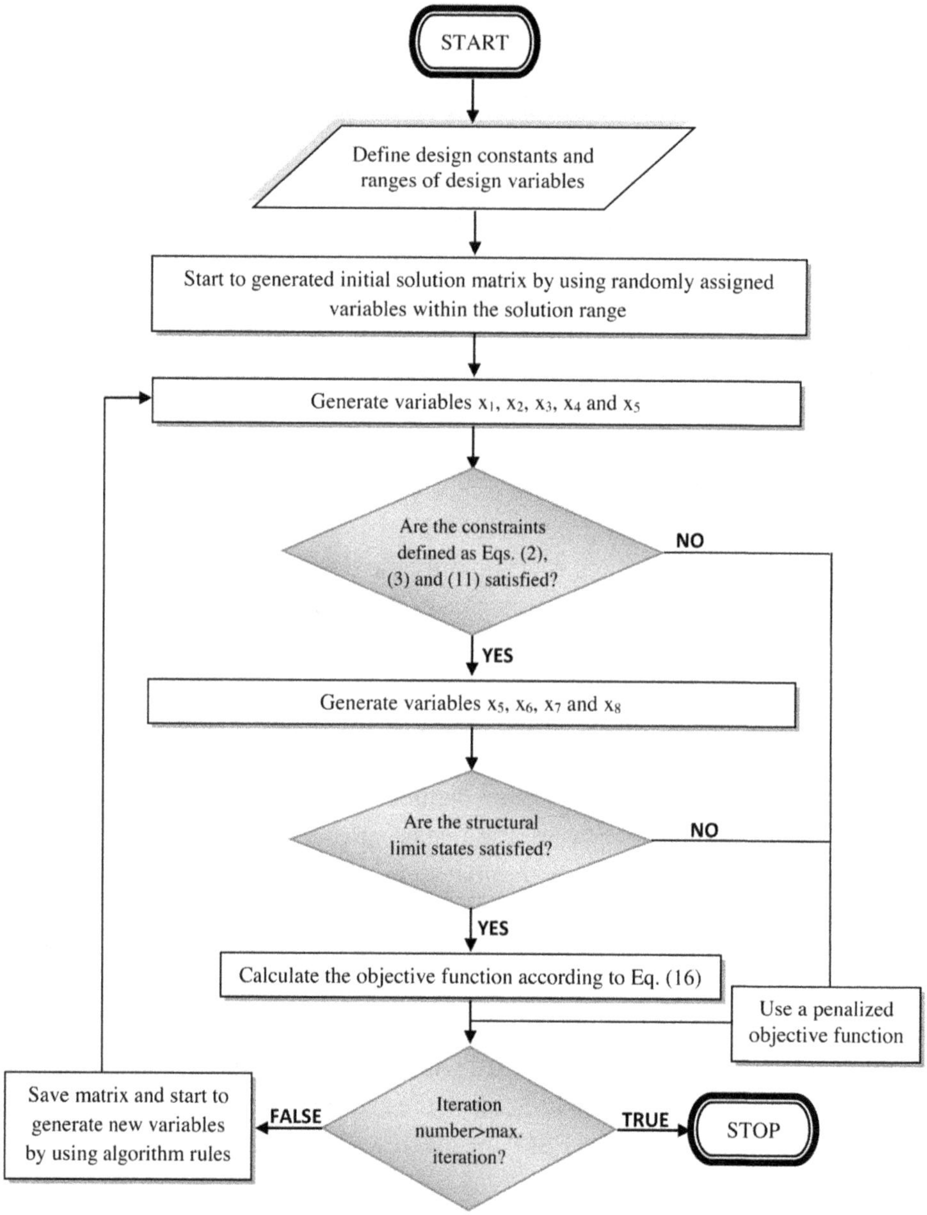

Table 1. Design constants and ranges of design variables

Definition	Symbol	Unit	Value
Yield strength of steel	f_y	MPa	420
Compressive strength of concrete	f'_c	MPa	25
Concrete cover	c_c	mm	100
Max. aggregate diameter	D_{max}	mm	16
Elasticity modulus of steel	E_s	GPa	200
Specific gravity of steel	γ_s	t/m^3	7.86
Specific gravity of concrete	γ_c	kN/m^3	23.5
Cost of concrete per m^3	C_c	unit/m^3	40
Cost of steel per ton	C_s	unit/t	400
Internal friction angle of soil	ϕ'	°	35
Unit weight of base soil	γ_B	kN/m^3	18.5
Poisson ratio of soil	ν	-	0.3
Modulus of elasticity of soil	E	MPa	50
Maximum allowable settlement	δ_{max}	mm	25
Factor of safety	FS	-	3.0
Minimum footing thickness	h_{min}	m	0.25
Column breadth in two direction	b/h	mm/mm	500/500
Range of width of footing	B	m	2.0-5.0
Range of length of footing	L	m	2.0-5.0
Range of height of footing	H	m	h_{min}-1.0
Range of diameter of reinforcement bars of two direction	ϕ	mm	16-24
Range of distance between reinforcement bars	s	mm	5ϕ-250

(Nigdeli et al., 2018)

Table 2. The loading cases for the optimization example

	P (kN)		M_x (kNm)		M_y (kNm)	
	P_G	P_Q	M_{xG}	M_{xQ}	M_{yG}	M_{yQ}
Case 1	750	500	300	200	300	200
Case 2	750	500	300	200	400	300
Case 3	750	500	400	300	300	200
Case 4	750	500	400	300	400	300
Case 5	1000	750	600	400	600	400

(Nigdeli et al., 2018)

Table 3. The optimum results and comparison of different algorithm (Case 1-4)

		Case Number			
		1	2	3	4
Optimum design variables	B (m)	2.75	2.60	3.20	3.05
	L (m)	2.65	3.20	2.60	3.05
	H (m)	1.00	1.00	1.00	1.00
	e_x (m)	0.75	0.75	0.75	0.90
	e_y (m)	0.55	0.75	0.75	0.80
	ϕ_x (mm)	16	16	20	16
	S_x (mm)	150	140	230	140
	ϕ_y (mm)	16	20	16	20
	S_y (mm)	160	230	140	210
Best Cost (unit)		217.3235	252.2302	252.2302	285.4123
DE	Av. Cost (unit)	218.5794	253.3985	253.3545	287.1336
	Sta. Dev. (unit)	0.3229	0.3268	0.4763	0.9942
	Analyses for optimum	223700	118500	110300	93200
PSO	Av. Cost (unit)	220.8968	254.2458	254.4545	288.1056
	Sta. Dev. (unit)	1.9426	1.5769	1.5691	1.5778
	Analyses for optimum	358600	345750	152600	658400
HS	Av. Cost (unit)	218.3516	253.2736	253.4752	287.3481
	Sta. Dev. (unit)	0.4723	0.4146	0.4812	1.0326
	Analyses for optimum	901100	301005	545795	602117
FPA	Av. Cost (unit)	218.2720	253.1640	253.1547	286.5305
	Sta. Dev. (unit)	0.4852	0.6284	0.5691	1.1384
	Analyses for optimum	221800	116850	145150	98150
TLBO	Av. Cost (unit)	219.1628	253.7319	253.4578	286.6425
	Sta. Dev. (unit)	1.4007	0.8896	0.9886	1.6982
	Analyses for optimum	242200	276400	381700	126100

(Nigdeli et al., 2018)

Table 4. t_{stat} values for the algorithms and cases

	Case No			
	1	2	3	4
DE	-50.7478	-18.3332	-16.3304	-6.9131
HS	-38.1056	-16.5810	-14.3904	-5.1872
FPA	-38.2525	-12.1729	-16.1500	-9.7835
TLBO	-8.75364	-4.08478	-7.12900	-6.0921

(Nigdeli et al., 2018)

The optimum results of Case 5 are given as Table 5. Three algorithms, namely; FPA, TLBO and PSO are effective to find the best results, but PSO has a big standard deviation value for multiple runs. These three algorithms have the same results, but it can be seen differently. Since the loading of case 5 is symmetrical, the inverse optimum solutions of x and y directions is found for FPA, TLBO and PSO.

Optimum Design of RC Spread Footings for Columns Under Uniaxial Flexural Moment

In this chapter, an optimum design code employing HS was given in the appendix. The code is written in Matlab (2018), and the design constraints are checked according to ACI-318. To eliminate the complexity of the code, several simplifications are done according to the presented methodology for biaxial loading. In the code, only flexural moment in one direction is used. The eccentricity of column is not considered, and the columns are mounted at the midpoint of footing. Also, the cross-sections of footing are rectangular. The main optimization code is given as Appendix 1. A sub-code is used for calculation. This code is given as Appendix 2, and the name of the code is "calculations_single_bending".

Table 5. The optimum results and comparison of different algorithm (Case 5)

Optimum Design Variables	DE	PSO	HS	FPA	TLBO
B (m)	4.15	4.15	4.20	4.15	4.15
B (m)	4.15	4.15	4.15	4.15	4.15
H (m)	1.00	1.00	1.00	1.00	1.00
e_x (m)	1.00	1.00	1.10	0.95	1.00
e_y (m)	0.95	0.95	0.85	1.00	0.95
ϕ_x (mm)	16	22	18	18	22
S_x (mm)	110	210	140	150	210
ϕ_y (mm)	20	18	18	22	18
S_y (mm)	180	150	150	210	150
Best Cost (unit)	549.7689	549.2428	554.2151	549.2428	549.2428
Av. Cost (unit)	554.0413	602.9062	556.9158	553.6568	554.7540
Sta. Dev. (unit)	0.8603	69.08	2.33	2.5674	13.29
Analyses for optimum	474800	935900	846801	712050	394700

(Nigdeli et al., 2018)

REFERENCES

ACI-318. (2011). Building code requirements for structural concrete and commentary, metric version, *American Concrete Institute*.

Geem, Z. W., Kim, J. H., & Loganathan, G. V. (2001). A new heuristic optimization algorithm: Harmony search. *Simulation*, *76*(2), 60–68. doi:10.1177/003754970107600201

Kennedy, J., & Eberhart, R. C. 1995. Particle swarm optimization. *Proceedings of IEEE International Conference on Neural Networks No. IV*, November 27-December 1, pp. 1942–1948, Perth Australia. 10.1109/ICNN.1995.488968

Mathworks. (2018). MATLAB R2018a. The MathWorks Inc., Natick, MA.

Nigdeli, S. M., Bekdaş, G., & Yang, X. S. (2018). Metaheuristic optimization of reinforced concrete footings. *KSCE Journal of Civil Engineering*, *22*(11), 4555–4563. doi:10.100712205-018-2010-6

Poulos, H. G., & Davis, E. H. (1974). *Elastic solutions for soil and rock mechanics (No. BOOK)*. John Wiley.

Rao, R. V., Savsani, V. J., & Vakharia, D. P. (2011). Teaching–learning-based optimization: A novel method for constrained mechanical design optimization problems. *Computer Aided Design*, *43*(3), 303–315. doi:10.1016/j.cad.2010.12.015

Storn, R., & Price, K. (1997). Differential evolution–a simple and efficient heuristic for global optimization over continuous spaces. *Journal of Global Optimization*, *11*(4), 341–359. doi:10.1023/A:1008202821328

Whitman, R. V., & Richart, F. E. (1967). Design procedures for dynamically loaded foundations. *Journal of the Soil Mechanics and Foundations Division*, *93*(6), 169–193.

Yang, X. S. (2012). Flower pollination algorithm for global optimization. In *International conference on unconventional computing and natural computation* (pp. 240-249). Berlin, Germany: Springer. 10.1007/978-3-642-32894-7_27

APPENDIX 1

```
%DESIGN OPTIMIZATION OF REINFORCED CONCRETE FOOTINGS
clear all
clc
%1.ENTER DATA OF PROBLEM
fc_prime=16.7;        % specified compressive strength of
concrete,[Mpa]
fy=365;               % specified yield strength of
reinforcement,[Mpa]
gama_steel=7.86;      % unit volume weight of stell, [t/m^3]
gama_conc=23.5;       % unit weight of concrete, [kN/m^3]
gama_soil=18.5;       % unit weight of base soil, [kN/m^3]
frict_angle=35;       % Internal friction angle of soil,
[degree]
Es=200000;            % modulus of elasticity of reinforcement,
[Mpa]
cc=60;                % net thickness of concrete cover, [mm]
eps_cu=0.003;         % Maximum usable strain at extreme
concrete compression fiber
bx=600;               % width of column cross-section in
direction of moment, [mm]
by=600;               % width of column cross-section
perpendicular to the moment direction, [mm]
Mu=234.4*10^6;        % factored moment at column section in Bx
direction, [N*mm],
Pu=1502.2*10^3;       % factored axial force normal to column
cross section, [N]
Cc=40;                % unit cost of concrete [dollar/m^3]
Cs=400;               % unit cost of reinforcement [dollar/t]
FS=3;                 % factor of safety for bearing capacity
phi=1.4;              % strength reduction factor
smin=80;              % lower limit for spacing of reinforcement
smax=250;             % upper limit for spacing of reinforcement
% factor relating depth of equivalent rectangular compressive
stress block to neutral axis depth, ACI318 Section 10.2.7.3
if fc_prime<=28
   beta1=0.85;
end
if fc_prime>28
   beta1=0.85-0.05*(fc_prime-28)/7;
end
if beta1<0.65
   beta1=0.65;
end
% User defined algorithm parameters
pn=20;          % population number
```

```
maxiter=20000; % maximum iteration number
HMCR=0.5;       % harmony memory considering rate
PAR=0.2;        % pitch adjusting rate
% Ultimate limits of design variables
Hmin=250;       % lower limit for footing thickness
Hmax=1000;       % upper limit for footing thickness
Bmin=2000;      % lower limit for width of footing
Bmax=6000;      % upper limit for width of footing
%2. GENERATING OF INITIAL SOLUTION MATRIX
% Bx   : dimension of the footing in the direction of moment,
[mm]
% By   : dimension of the footing perpendicular to B, [mm]
% H    : footing thickness, [mm]
% rand : a function generating uniformly distributed
pseudorandom numbers in MATLAB
for i=1:pn % During the this loop, candidate solution vectors
including design variables is generated up to pn.
          % End of the this loop, candidate solution vectors
provided design constraint is stored in a matrix.
    % 2.1. DEFINITION OF DESING VARIABLES RANDOMLY
    H=Hmin+rand*(Hmax-Hmin);                      % Generation of
H between ultimate limits and randomly
    Bx=Bmin+rand*(Bmax-Bmin);                     % Generation of
Bx between ultimate limits and randomly
    By=Bmin+rand*(Bmax-Bmin);                     % Generation of
By between ultimate limits and randomly
    % 2.2. OTHER CALCULATIONS
    Wf=Bx*By*H*(gama_conc/(10^6)); % weight of footing
    % 2.2.2. Determination of maximum end minimum applied
bearing pressures on the base of footing:
    % Calculations in direction of Bx
    q1=((Pu+Wf)/(Bx*By))+(6*abs(Mu)/(By*Bx^2));
    q2=((Pu+Wf)/(Bx*By))-(6*abs(Mu)/(By*Bx^2));
    qmax=max(q1,q2); %maximum bearing pressure
    qmin=min(q1,q2); %minimum bearing pressure
    % Calculations in direction of By
    % NOTE: There is uniform pressure in the By direction for
the unit lenth due to single direction moment.
    %       Thus average pressure (qave) is taken into
consideration.
    qave=(qmax+qmin)*0.5;
    % 2.2.3. Calculation of monent and shear foces at the
critical sections or face of column (ACI318 Section 15.4.2)
    % NOTE: qA1, qA2 are pressures on the column face for A1-A1
and A2-A2 sections respectively.
    %       VA1, VA2 are shear forces on the column face for
A1-A1 and A2-A2 sections respectively.
    %       MA1, MA2 are momnets on the column face for A1-A1
and A2-A2 sections respectively.
```

```
    if Mu>0% if moment acts in counterclockwise:
       qA1=qmin+((qmax-qmin)/Bx)*((Bx+bx)*0.5);
       qA2=qmin+((qmax-qmin)/Bx)*((Bx-bx)*0.5);
       VA1=By*(qmax+qA1)*(Bx-bx)*0.25;
       VA2=By*(qmin+qA2)*(Bx-bx)*0.25;
       MA1=By*(((Bx-bx)^2)/24)*(2*qmax+qA1);
       MA2=By*(((Bx-bx)^2)/24)*(2*qmin+qA2);
    else    % if moment acts in clockwise:
       qA1=qmin+((qmax-qmin)/Bx)*((Bx-bx)*0.5);
       qA2=qmin+((qmax-qmin)/Bx)*((Bx+bx)*0.5);
       VA1=By*(qmin+qA1)*(Bx-bx)*0.25;
       VA2=By*(qmax+qA2)*(Bx-bx)*0.25;
       MA1=By*(((Bx-bx)^2)/24)*(2*qmin+qA1);
       MA2=By*(((Bx-bx)^2)/24)*(2*qmax+qA2);
    end
    % Design forces:
    VA=max(VA1,VA2);          % shear force for A-A section:
    MA=max(MA1,MA2);          % moment for A-A section:
    VD=Bx*qave*(By-by)*0.5; % shear force for D-D section:
    MD=VD*(By-by)*0.25;       % moment for D-D section:
    % 2.2.4. Determination of diameters of reinforcement bars,
spacings between reinforment bars and reinforcement areas for
one meter:
    R=16:2:40; % range of diameters of reinforcement bars
    for m=1:length(R)
        % Distances from extreme compression fiber to centroid
of longitudinal tension reinforcement:
        d=H-cc-0.5*R(m);
        % Depth of equivalent rectangular stress blocks as
defined in ACI318-10.2.7.1 [mm]:
        a_A=d-sqrt(d^2-2*MA/(0.85*fc_prime*1000)); % for A-A
section
        a_D=d-sqrt(d^2-2*MD/(0.85*fc_prime*1000)); % for D-D
section
        % In this part, if d^2-2*MA/(0.85*fc_prime*1000)<0 or
d^2-2*MD/(0.85*fc_prime*1000)<0,
        % lower limits of requared reinforcement areas (As_Amin
and As_Dmin), provided reinforcement areas (As_provided)
        %  and spacing between the reinforcements (s_A and s_D)
are penaltized with high value like 10^16,
        % since this solution will no be physically possible
        if d^2-2*MA/(0.85*fc_prime*1000)<0 || d^2-2*MD/
(0.85*fc_prime*1000)<0
           % for A-A section
           As_Amin(m)=10^16;
           s_A(m)=10^16;
           As_Aprovided(m)=10^16;
           % for D-D section
           As_Dmin(m)=10^16;
```

```
            s_D(m)=10^16;
            As_Dprovided(m)=10^16;
            % NOTE: In this part, the main purpose is to prevent
the selection of
            % provided reinforcement areas depending on 'if-
end'loop condition.
            % In this 'if-end'loop, spacing and lower limits of
areas must also
            % be assigned values. Otherwise, the program cannot
continue. In
            % this case, penalty value can be used.
        else %(d^2-2*MA/(0.85*fc_prime*1000)>0 and d^2-2*MD/
(0.85*fc_prime*1000)>0)
            % Requared reinforcement areas:
            As_Ar=0.85*fc_prime*a_A*1000/fy;
            As_Dr=0.85*fc_prime*a_D*1000/fy;
            % Lower limits of required reinforcement areas:
            As_Amin(m)=max(As_Ar,0.005*(H*1000-As_Ar));
            As_Dmin(m)=max(As_Dr,0.005*(H*1000-As_Dr));
            % Spacing between the reinforcements for different
rebars diameters, [mm]:
            s_A(m)=fix(1000/(As_Amin(m)/(0.25*pi*(R(m))^2)));
            s_D(m)=fix(1000/(As_Dmin(m)/(0.25*pi*(R(m))^2)));
            % Ultimate limit of spacing must not be exceeded:
            if s_A(m)>smax
               s_A(m)=smax;
            end
            if s_D(m)>smax
               s_D(m)=smax;
            end
            % Providing reinforcement areas considering assigned
rebar diameter and calculated rebar spacing [mm^2]:
            As_Aprovided(m)=(1000/s_A(m))*(0.25*pi*(R(m))^2);
            As_Dprovided(m)=(1000/s_D(m))*(0.25*pi*(R(m))^2);
            %================================================
            % For values that do not provide lower limit of the
rebar spacing, provided reinforcement areas are penalized.
            % Thus, the selection of improper provided
reinforcement is prevented:
            if s_A(m)<smin || s_D(m)<smin
               As_Aprovided(m)=10^16;
               As_Dprovided(m)=10^16;
            end
        end
    end
    % Determination of the area closest to the required
reinforcement area
    % Calculatoins for reinforcement bars placed in the Bx
direction:
```

```
    [min_value,column_number]=min(As_Aprovided);
    As_Aselected=min_value; % area of reinforcement bars, [mm]
    R_A=R(column_number);        % diameter of reinforcement
bars, [mm]
    s_A=s_A(column_number);       % spacing between the
reinforcements, [mm]
    % Calculatoins for reinforcement bars placed in the Bx
direction:
    [min_value,column_number]=min(As_Dprovided);
    As_Dselected=min_value; % area of reinforcement bars, [mm]
    R_D=R(column_number);        % diameter of reinforcement
bars, [mm]
    s_D=s_D(column_number);       % spacing between the
reinforcements, [mm]
    %2.2.5. Ultimate bearing capacity of soil (qult):
    Nq=exp(pi*tand(frict_angle))*(tand(45+0.5*frict_angle))^2;
    Ngama=2*(Nq+1)*tand(frict_angle);
    Fqs=1+(min(Bx,By)/max(Bx,By))*tand(frict_angle);
    Fgama_s=1-0.4*(min(Bx,By)/max(Bx,By));
    if H>min(Bx,By)
        Fqd=1+2*(tand(frict_angle))*((1-sind(frict_
angle))^2)*atand(H/min(Bx,By));
    else
        Fqd=1+2*(tand(frict_angle))*((1-sind(frict_
angle))^2)*H/min(Bx,By);
    end
    Fgama_d=1;
    qult=gama_soil*H*Nq*Fqs*Fqd+0.5*gama_soil*min(Bx,By)*Ngama*
Fgama_s*Fgama_d;
    % 2.2.6. Shear capacity of concrete
    dA=H-cc-0.5*R_A;
    dD=H-cc-0.5*R_D;
    Vc_A=0.17*sqrt(fc_prime)*By*dA;
    Vc_D=0.17*sqrt(fc_prime)*Bx*dD;
    % 2.2.7. Punching forces (ACI318-11.12.2.1)
   q1u=qmin+((qmax-qmin)/Bx)*((Bx-bx)*0.5-dA);
   q2u=qmin+((qmax-qmin)/Bx)*((Bx+bx)*0.5+dA);
   Vz=0.5*(q1+q2)*Bx*By-0.5*(q1u+q2u)*(bx+dA)*(by+dD);
    Betac=max(bx,by)/min(bx,by);
    b0=2*(bx+by+dA+dD); %punching circumference
    dave=(dA+dD)/2;
    Vcz1=(1+(2/Betac))*(fc_prime^0.5*b0*dave/6);
    Vcz2=(2+(40*dave/b0))*(fc_prime^0.5*b0*dave/12);
    Vcz3=(fc_prime^0.5*b0*dave/3);
    Vcz=0.75*(min(min(Vcz1,Vcz2),Vcz3));
    % 2.3. OBJECTIVE FUNCTON
    % NOTE: The values of objective function is calculated in
every loop and through minimum value of objective function
    % the  candidate solution vector including values of
```

```
candidate optimum design variables can be found.
    %Total material cost of footing:
    F_obj=Bx*By*H*Cc/(10^9)+gama_steel*(((Bx-cc))*As_
Aselected+((By-cc))*As_Dselected)*Cs/(10^9);
    % 2.4. PART OF PENALIZATION
    % NOTE: Candidate solution vectors that do not meet design
constraints are penalized with a very large objective function
value.
    % With this penaltization, selection of a solution vector
including improper value or values of design variables are
prevented.
    if qmin<0 || qult/qmax<FS || Vc_A<VA || Vc_D<VD || Vz>=Vcz
%||...
       %d^2-2*MA/(0.85*fc_prime*1000)<0 || d^2-2*MD/(0.85*fc_
prime*1000)<0 ||...
       %s_A<smin || s_D<smin
       %Objective function:
       F_obj=10^16;  % Penalization with high value
    end
    % 2.5. INITIAL SOLUTION MATRIX
    % NOTE: End of the each loop, a candidate solutution vector
including randomly generated design variables is stored in the
matrix named OPT
    % Thus each vector in matrix is candidate solution vector.
    OPT(1,i)=H;
    OPT(2,i)=Bx;
    OPT(3,i)=By;
    OPT(4,i)=R_A;
    OPT(5,i)=s_A;
    OPT(6,i)=As_Aselected;
    OPT(7,i)=MA;
    OPT(8,i)=R_D;
    OPT(9,i)=s_D;
    OPT(10,i)=As_Dselected;
    OPT(11,i)=MD;
    OPT(12,i)=F_obj;
end
% 3. GENERATING OF NEW SOLUTION MATRIX VIA HS
for iter=1:maxiter
    % 3.1. SELECTION OF GLOBAL or LOCAL SEARCH ACCORDING TO
HARMONY MEMORY SIZE CONSIDERING RATE
    % 3.1.1. GLOBAL PHASE
    if rand()<HMCR
       % 3.1.1.1. DEFINATION OF DESING VARIABLES
       H=Hmin+rand*(Hmax-Hmin);
       Bx=Bmin+rand*(Bmax-Bmin);
       By=Bmin+rand*(Bmax-Bmin);
       % for other calculations can be called subcode named
"calculations_for_footing"
```

```
        calculations_single_bending
    else
        % 3.1.2. LOCAL PHASE
        % In this phase new value of design variables are
generated according to randomly selected candidate solution
vector.
        % A random number is determined between zero and pn:
        v1=ceil(rand*pn);
        % 3.1.2.1. DEFINATION OF DESING VARIABLES
        H=OPT(1,v1)+(rand-0.5)*PAR*(Hmax-Hmin);
        Bx=OPT(2,v1)+(rand-0.5)*PAR*(Bmax-Bmin);
        By=OPT(3,v1)+(rand-0.5)*PAR*(Bmax-Bmin);
        % for other calculations can be called subcode named
"calculations_for_footing"
        calculations_single_bending
    end
[mi,ni]=min(OPT(12,:));
Fmin(iter,1)=mi;
iter
mi
end
```

APPENDIX 2

```
% 1. CONTROL OF ULTIMATE LIMITS
if H<Hmin
   H=Hmin;
end
if H>Hmax
   H=Hmax;
end
if Bx<Bmin
   Bx=Bmin;
end
if Bx>Bmax
   Bx=Bmax;
end
if By<Bmin
   By=Bmin;
end
if By>Bmax
   By=Bmax;
end
% 2. OTHER CALCULATIONS
Wf=Bx*By*H*(gama_conc/(10^6)); % weight of footing
% 2.2.2. Determination of maximum end minimum applied
bearing pressures on the base of footing:
% Calculations in direction of Bx
q1=((Pu+Wf)/(Bx*By))+(6*abs(Mu)/(By*Bx^2));
q2=((Pu+Wf)/(Bx*By))-(6*abs(Mu)/(By*Bx^2));
qmax=max(q1,q2); %maximum bearing pressure
qmin=min(q1,q2); %minimum bearing pressure
% Calculations in direction of By
% NOTE: There is uniform pressure in the By direction for
the unit lenth due to single direction moment.
%       Thus average pressure (qave) can be taken
consideration.
qave=(qmax+qmin)*0.5;
% 2.1. Calculation of monent and shear foces at the
critical sections or face of column (ACI318 Section 15.4.2)
% NOTE: qA1, qA2 are pressures on the column face for A1-A1
and A2-A2 sections respectively.
%       VA1, VA2 are shear forces on the column face for
A1-A1 and A2-A2 sections respectively.
%       MA1, MA2 are momnets on the column face for A1-A1
and A2-A2 sections respectively.
if Mu>0% if moment acts in counterclockwise:
   qA1=qmin+((qmax-qmin)/Bx)*((Bx+bx)*0.5);
   qA2=qmin+((qmax-qmin)/Bx)*((Bx-bx)*0.5);
```

```
        VA1=By*(qmax+qA1)*(Bx-bx)*0.25;
        VA2=By*(qmin+qA2)*(Bx-bx)*0.25;
        MA1=By*(((Bx-bx)^2)/24)*(2*qmax+qA1);
        MA2=By*(((Bx-bx)^2)/24)*(2*qmin+qA2);
    else    % if moment acts in clockwise:
        qA1=qmin+((qmax-qmin)/Bx)*((Bx-bx)*0.5);
        qA2=qmin+((qmax-qmin)/Bx)*((Bx+bx)*0.5);
        VA1=By*(qmin+qA1)*(Bx-bx)*0.25;
        VA2=By*(qmax+qA2)*(Bx-bx)*0.25;
        MA1=By*(((Bx-bx)^2)/24)*(2*qmin+qA1);
        MA2=By*(((Bx-bx)^2)/24)*(2*qmax+qA2);
    end
    % Design forces:
    VA=max(VA1,VA2);          % shear force for A-A section:
    MA=max(MA1,MA2);          % moment for A-A section:
    VD=Bx*qave*(By-by)*0.5;   % shear force for D-D section:
    MD=VD*(By-by)*0.25;       % moment for D-D section:
    % 2.2. Determination of diameters of reinforcement bars,
spacings between reinforment bars and reinforcement areas for
one meter:
    R=16:2:40; % range of diameters of reinforcement bars
    for m=1:length(R)
        % Distances from extreme compression fiber to centroid
of longitudinal tension reinforcement:
        d=H-cc-0.5*R(m);
        % Depth of equivalent rectangular stress blocks as
defined in ACI318-10.2.7.1 [mm]:
        a_A=d-sqrt(d^2-2*MA/(0.85*fc_prime*1000)); % for A-A
section
        a_D=d-sqrt(d^2-2*MD/(0.85*fc_prime*1000)); % for D-D
section
        % In this part, if d^2-2*MA/(0.85*fc_prime*1000)<0 or
d^2-2*MD/(0.85*fc_prime*1000)<0,
        % lower limits of requared reinforcement areas (As_Amin
and As_Dmin), provided reinforcement areas (As_provided)
        %  and spacing between the reinforcements (s_A and s_D)
are penaltized with high value like 10^16,
        % since this solution will no be physically possible
        if d^2-2*MA/(0.85*fc_prime*1000)<0 || d^2-2*MD/
(0.85*fc_prime*1000)<0
            % for A-A section
            As_Amin(m)=10^16;
            s_A(m)=10^16;
            As_Aprovided(m)=10^16;
            % for D-D section
            As_Dmin(m)=10^16;
            s_D(m)=10^16;
            As_Dprovided(m)=10^16;
            % NOTE: In this part, the main purpose is to prevent
```

```
the selection of
            % provided reinforcement areas depending on 'if-
end'loop condition.
            % In this 'if-end'loop, spacing and lower limits of
areas must also
            % be assigned values. Otherwise, the program cannot
continue. In
            % this case, penalty value can be used.
        else %(d^2-2*MA/(0.85*fc_prime*1000)>0 and d^2-2*MD/
(0.85*fc_prime*1000)>0)
            % Requared reinforcement areas:
            As_Ar=0.85*fc_prime*a_A*1000/fy;
            As_Dr=0.85*fc_prime*a_D*1000/fy;
            % Lower limits of requared reinforcement areas:
            As_Amin(m)=max(As_Ar,0.005*(H*1000-As_Ar));
            As_Dmin(m)=max(As_Dr,0.005*(H*1000-As_Dr));
            % Spacing between the reinforcements for different
rebars diameters, [mm]:
            s_A(m)=fix(1000/(As_Amin(m)/(0.25*pi*(R(m))^2)));
            s_D(m)=fix(1000/(As_Dmin(m)/(0.25*pi*(R(m))^2)));
            % Ultimate limit of spacing must not be exceeded:
            if s_A(m)>smax
               s_A(m)=smax;
            end
            if s_D(m)>smax
               s_D(m)=smax;
            end
            % Provided reinforcement areas considering assigned
rebar diameter and calculated rebar spacing [mm^2]:
            As_Aprovided(m)=(1000/s_A(m))*(0.25*pi*(R(m))^2);
            As_Dprovided(m)=(1000/s_D(m))*(0.25*pi*(R(m))^2);
            if s_A(m)<smin || s_D(m)<smin
               As_Aprovided(m)=10^16;
               As_Dprovided(m)=10^16;
            end
        end
    end
    % Determination of the area closest to the required
reinforcement area
    % Calculatoins for reinforcement bars placed in the Bx
direction:
    [min_value,column_number]=min(As_Aprovided);
    As_Aselected=min_value; % area of reinforcement bars, [mm]
    R_A=R(column_number);        % diameter of reinforcement
bars, [mm]
    s_A=s_A(column_number);        % spacing between the
reinforcements, [mm]
    % Calculatoins for reinforcement bars placed in the Bx
direction:
```

```
    [min_value,column_number]=min(As_Dprovided);
    As_Dselected=min_value; % area of reinforcement bars, [mm]
    R_D=R(column_number);       % diameter of reinforcement
bars, [mm]
    s_D=s_D(column_number);      % spacing between the
reinforcements, [mm]
    %2.2.5. Ultimate bearing capacity of soil (qult):
    Nq=exp(pi*tand(frict_angle))*(tand(45+0.5*frict_angle))^2;
    Ngama=2*(Nq+1)*tand(frict_angle);
    Fqs=1+(min(Bx,By)/max(Bx,By))*tand(frict_angle);
    Fgama_s=1-0.4*(min(Bx,By)/max(Bx,By));
    if H>min(Bx,By)
        Fqd=1+2*(tand(frict_angle))*((1-sind(frict_
angle))^2)*atand(H/min(Bx,By));
    else
        Fqd=1+2*(tand(frict_angle))*((1-sind(frict_
angle))^2)*H/min(Bx,By);
    end
    Fgama_d=1;
    qult=gama_soil*H*Nq*Fqs*Fqd+0.5*gama_soil*min(Bx,By)*Ngama*
Fgama_s*Fgama_d;
    % 2.3. Shear capacity of concrete
    dA=H-cc-0.5*R_A;
    dD=H-cc-0.5*R_D;
    Vc_A=0.17*sqrt(fc_prime)*By*dA;
    Vc_D=0.17*sqrt(fc_prime)*Bx*dD;
    % 2.4. Punching forces (ACI318 Section 11.12.2.1)
   q1u=qmin+((qmax-qmin)/Bx)*((Bx-bx)*0.5-dA);
   q2u=qmin+((qmax-qmin)/Bx)*((Bx+bx)*0.5+dA);
   Vz=0.5*(q1+q2)*Bx*By-0.5*(q1u+q2u)*(bx+dA)*(by+dD);
    Betac=max(bx,by)/min(bx,by);
    b0=2*(bx+by+dA+dD); %punching circumference
    dave=(dA+dD)/2;
    Vcz1=(1+(2/Betac))*(fc_prime^0.5*b0*dave/6);
    Vcz2=(2+(40*dave/b0))*(fc_prime^0.5*b0*dave/12);
    Vcz3=(fc_prime^0.5*b0*dave/3);
    Vcz=0.75*(min(min(Vcz1,Vcz2),Vcz3));
    % 3. OBJECTIVE FUNCTON
    % NOTE: The values of objective function is calculated in
every loop and through minimum value of objective function
    % the  candidate solution vector including values of
candidate optimum design variables can be found.
    %Total material cost of footing:
    F_obj=Bx*By*H*Cc/(10^9)+gama_steel*(((Bx-cc))*As_
Aselected+((By-cc))*As_Dselected)*Cs/(10^9);
    % 4. PART OF PENALIZATION
    % NOTE: Candidate solution vectors that do not meet design
constraints are penalized with a very large objective function
value.
```

```
    % With this penaltization, selection of a solution vector including improper value or values of design variables is prevented.
    if qmin<0 || qult/qmax<FS || Vc_A<VA || Vc_D<VD || Vz>=Vcz %||...
       %d^2-2*MA/(0.85*fc_prime*1000)<0 || d^2-2*MD/(0.85*fc_prime*1000)<0 ||...
       %s_A<smin || s_D<smin
       %Objective function:
       F_obj=10^16; % Penalization with high value
    end
    % 5. New SOLUTION MATRIX
    % NOTE: End of the each loop, a candidate solututition vector including randomly generated design variables stored in the matrix named OPT
    % Thus each vector in matrix is candidate solution vector.
    OPT1(1,1)=H;
    OPT1(2,1)=Bx;
    OPT1(3,1)=By;
    OPT1(4,1)=R_A;
    OPT1(5,1)=s_A;
    OPT1(6,1)=As_Aselected;
    OPT1(7,1)=MA;
    OPT1(8,1)=R_D;
    OPT1(9,1)=s_D;
    OPT1(10,1)=As_Dselected;
    OPT1(11,1)=MD;
    OPT1(12,1)=F_obj;
   % 6.PART OF COMPARISON
     % Determination of vector number giving worst or maximum value of objective function
     [ke,te]=max(OPT(12,:));
     % If new solution vector (OPT1) provide better solution or minimum cost replaced with existing worst solution vector:
     if OPT(12,te)>OPT1(12,1) % Comparing of values of objective functions
        OPT(:,te)=OPT1;       % Updating of matrix OPT
     end
```

Chapter 6
Metaheuristic Optimization of Reinforced Concrete Frames:
Optimization of RC Frames

ABSTRACT

In this chapter, an optimum design methodology for reinforced concrete (RC) frames are presented. In the optimum design of frames, both beams and columns are optimized. In addition to that, internal forces can be modified according to rigidity of members for statically indetermined frames. In the presented methodology, the optimization of RC frame is done according to dynamic seismic loads and the design is done according to time-history analysis. As a metaheuristic algorithm, a modified harmony search is used, and the design constraints are provided according to ACI 318: Building code requirements for structural concrete. The optimum results of two span-two story symmetric RC frames and three span-three story RC frame are presented.

INTRODUCTION

In the previous chapters, the optimum designs of several reinforced concrete (RC) members are presented. By combining methodologies, it is possible to optimize a full structure, but by the selection of different cross-sectional dimension, the rigidity changes of statically indetermined structures affect the values of internal forces. For that reason, the dimension design variables must be selected before the determination of internal forces.

DOI: 10.4018/978-1-7998-2664-4.ch006

In this chapter, an optimum design methodology for RC frames (Nigdeli & Bekdaş, 2016; Bekdaş & Nigdeli, 2017) is presented by employing a modified harmony search algorithm (Geem, Kim, Loganathan, 2001). The methodology considers time-history analysis under earthquake excitations.

METHODOLOGY

A methodology for RC frame structures excited by static and dynamic forces is presented. The methodology employees a modified harmony search (HS) algorithm using additional random search iterations since the optimization of frame structures includes optimization of various numbers of beams and columns. In additional to the optimum design of beams and columns, spread footings positioned under the base level columns can be also optimized according to the methods given in Chapter 5.

As the interesting part of the presented methodology, dynamic forces resulting from earthquakes are considered via time history analyses of three chosen earthquake records. In all iterations of optimization methodology, all time history analyses are redone according to chosen candidate cross-section dimensions. After the time history analyses, the earthquake effects are distributed to all joints. Also, angular displacement of frames is considered under earthquake excitations.

Another important factor for an excellent optimum design of frames are to consider all unfavorable conditions of load combinations and existing of live loads in spans. Thus, an optimum solution for the most critical condition can be found.

In several methodologies, beam and column elements are not grouped and templates are given for rebar combinations and sections. In the presented methodology, rebar design properties are also optimized as separate design variables.

All calculations about structural state limits are checked according to ACI-318: Building code requirements for structural concrete by formulating the rules and theory as design constraints. The methodology was coded by using Matlab (2018). Time history analyses are done by modelling equations of motion of frame in Matlab Simulink.

Firstly, design constants such as all properties of the structural model, material properties, loadings, earthquake data, design variable ranges and algorithm parameters (harmony memory size, harmony memory considering rate and pitch adjusting rate) are defined. These properties include number of bays, number of stories, number of joints, boundary conditions of joints, coordinates of elements and the joints. Ranges of cross-section dimensions (breadth; b_w and height; h) and reinforcement sizes are defined. Loading conditions are defined as Live (L) and Dead (D) distributed loads except self-weight of beams and, columns are defined for all spans, because the self-weight of optimized members is calculated according to assigned dimensions. These loads can be entered as equally distributed, triangular distributed or trapezium distributed as seen in Fig. 1 according to the assumption used in transferring of loads from the slabs. The other design constants include clear cover (c_c), maximum aggregate diameter (D_{max}), compressive strength of concrete (f_c'), yield strength of steel (f_y), elasticity modulus of steel (E_s) and specific gravity of steel (γ_s), specific gravity of concrete (γ_c), the cost of concrete per m³ (C_c), the cost of steel per ton (C_s) are defined. The elasticity modulus of concrete (E_c) is defined with Eq. (1) according to compressive strength of concrete.

$$E_c = 4700\sqrt{f_c'} \tag{1}$$

After the definition of constant parameters, an initial harmony matrix including sets of candidate solutions are defined. For initial values, the values of design variables are randomly chosen within the range. A set of candidate solutions generates a harmony vector and the number of these vectors is equal to harmony memory size. The design variables are particularly assigned after several randomization iterations finished. In these iterations, several design constraints are checked, and computational effort of the methodology is saved. These stages are mainly; randomization of cross-sectional dimensions supporting design code requirements and rebar design with closest area to required area.

Firstly, cross-sectional dimensions of the beams are randomized by providing Eqs. 2 and 3. Randomizations repeated until these constraints are provided.

Figure 1. Equally distributed, triangular distributed and trapezium distributed loads

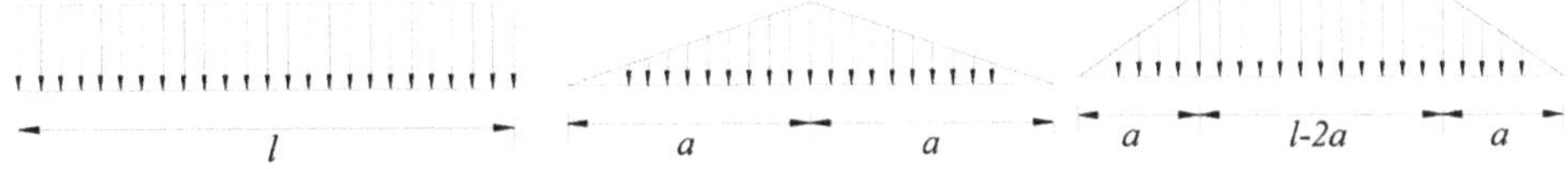

$$d < \frac{l}{4} \tag{2}$$

$$b_w \geq 0.3h \tag{3}$$

In Eq. (2), depth (d) of the beam is limited to 1/4 of clear length of the beam (*l*). Since the sizes of rebars are not known, depth of the beam is taken as 50 mm less than the height of the beam. In the further calculations, the exact value of d is calculated according to assigned rebar sizes for capacity checks and, this condition is also rechecked.

Then, cross-sectional dimensions of columns are randomized until Eq. (4) is provided. In Eq. (4), b is the breadth of the supporting column of a beam checked for this constraint, and d and b_w dimensions are for that beam.

$$b_w \leq b + \frac{3}{2}d \tag{4}$$

After all dimensions of beams and columns are defined, static and dynamic analyses can be done to provide internal forces. Static analysis of statically indetermined frame is done by using the stiffness method and dynamic analysis for earthquakes are done via time history analyses. For these analyses, mass and stiffness matrices of the system are needed to be constructed. Consistent mass (M) and stiffness (K) matrices of structure is constructed by combining element mass and stiffness matrices, and by eliminating fixed freedoms of supports. Mass (M_e) and stiffness (K_e) matrices of a structural member are given in Eqs. (5) and (6), respectively. In these matrices, g, E, *l*, I, A and n are gravity, elasticity modulus (taken as E_c), length, moment of inertia, cross-sectional area of the element and live load participation factor, respectively.

$$M_e = \frac{\gamma_c Al + (D + nL)(l - a)}{420g} \begin{bmatrix} 140 & 0 & 0 & 70 & 0 & 0 \\ 0 & 156 & 22l & 0 & 54 & -13l \\ 0 & 22l & 4l^2 & 0 & 13l & -3l^2 \\ 70 & 0 & 0 & 140 & 0 & 0 \\ 0 & 54 & 13l & 0 & 156 & -22l \\ 0 & -13l & -3l^2 & 0 & -22l & 4l^2 \end{bmatrix} \tag{5}$$

$$K_e = \begin{bmatrix} \frac{EA}{l} & 0 & 0 & -\frac{EA}{l} & 0 & 0 \\ 0 & \frac{12EI}{l^3} & \frac{6EI}{l^2} & 0 & -\frac{12EI}{l^3} & \frac{6EI}{l^2} \\ 0 & \frac{6EI}{l^2} & \frac{4EI}{l} & 0 & -\frac{6EI}{l^2} & \frac{2EI}{l} \\ -\frac{EA}{l} & 0 & 0 & \frac{EA}{l} & 0 & 0 \\ 0 & -\frac{12EI}{l^3} & -\frac{6EI}{l^2} & 0 & \frac{12EI}{l^3} & -\frac{6EI}{l^2} \\ 0 & \frac{6EI}{l^2} & \frac{2EI}{l} & 0 & -\frac{6EI}{l^2} & \frac{4EI}{l} \end{bmatrix} \tag{6}$$

The equation of structures subjected to ground acceleration ($\ddot{x}_g(t)$) is written given as Eq. (7).

$$M\ddot{x}(t) + C\dot{x}(t) + Kx(t) = -M\left\{1\right\}\ddot{x}_g(t) \tag{7}$$

x(t) is deflection vector including displacements and rotations. C is damping matrix and it is calculated by assuming an inherent damping ratio (5% for RC structures) for all modes. [1] is a vector of ones and size of this vector is equal to number of degrees of freedom of structure. First and second derivatives of x(t) with respect to time are respectively shown with $\dot{x}(t)$ and $\ddot{x}(t)$. The equation of motion is modelled with Matlab Simulink to carry out the time history analyses. The coupled equations of motions are separated into vibration modes and the results are obtained for all modes, and then, the results of all vibration modes are combined. The mode-shape matrix (Φ) can be used to transform the generalized coordinates; Y(t) into the geometric coordinates; x(t). By using this idea and premultiplying the transpose of the ith mode-shape vector; ϕ_i^T, the equation of motion takes the form given as Eq. (8).

$$\varphi_i^T M\Phi\ddot{Y}(t) + \varphi_i^T C\Phi\dot{Y}(t) + \varphi_i^T K\Phi Y(t) = -\varphi_i^T M\left\{1\right\}\ddot{x}_g(t) \tag{8}$$

Eq. (8) is written as Eq. (9), if the properties of mode shapes called orthogonality conditions given as Eqs. (10-15) are considered.

$$M_i\ddot{Y}(t) + C_i\dot{Y}(t) + K_iY(t) = -\varphi_i^T M\{1\}\ddot{x}_g(t) \tag{9}$$

$$M_i = \varphi_i^T \mathrm{M} \varphi_i \tag{10}$$

$$\varphi_n^T \mathrm{M} \varphi_m = 0, n \neq m \tag{11}$$

$$K_i = \varphi_i^T K \varphi_i \tag{12}$$

$$\varphi_n^T K \varphi_m = 0, n \neq m \tag{13}$$

$$C_i = \varphi_i^T C \varphi_i \tag{14}$$

$$\varphi_n^T C \varphi_m = 0, n \neq m \tag{15}$$

M_i, C_i, K_i and $Y_i(t)$ represents generalized mass, damping, stiffness and deflection of ith normal mode, respectively. For further modification of equation of motion of a vibration mode, natural frequency and damping ratio of ith normal mode represented by ω_i and ξ_i (Eqs. (16) and (17)) are used. By dividing Eq. (9) to M_i and substituting Eqs. (16) and (17), Eq. (18) is ready for modelling in Simulink as shown in block diagram given as Fig. 2.

$$w_i^2 = \frac{K_i}{M_i} \tag{16}$$

$$2\xi_i w_i = \frac{C_i}{M_i} \tag{17}$$

$$\ddot{Y}_i(t) + 2\xi_i w_i \dot{Y}_i(t) + w_i^2 Y_i(t) = -\varphi_i^T M / M_i \{1\} \ddot{x}_g(t) \tag{18}$$

Figure 2. Block diagram of Eq. (18)
(Nigdeli & Bekdaş, 2016)

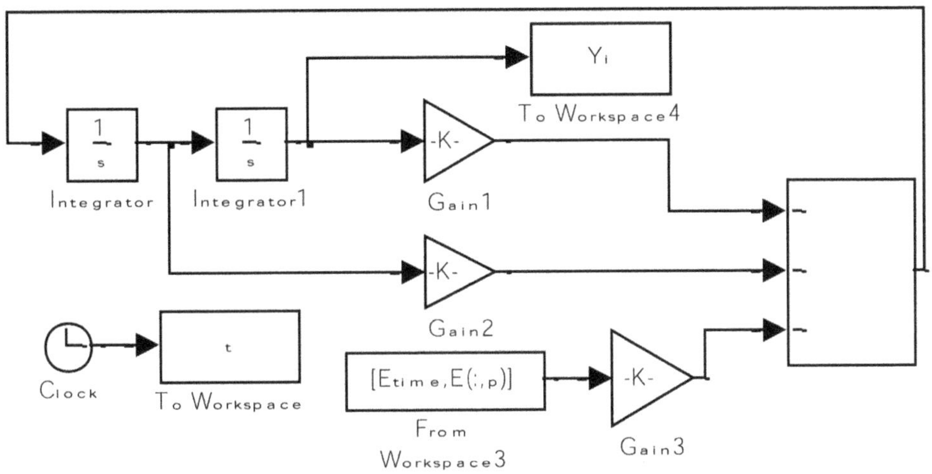

The analyses are conducted for all modes and the results are transformed into geometric coordinates. After solutions of deflections are found, internal forces are calculated by multiplying the stiffness matrix with deflections divided to the elastic response parameter (R). The load combinations given as Eqs. (19)-(21) are calculated for all combinations of live load distributions and all time lags of time history analysis in order to find the most critical situation for members. U and E represent total load and earthquake load, respectively.

$$U = 1.4D + 1.7L \tag{19}$$

$$U = 0.75(1.4D + 1.7L) \pm E \tag{20}$$

$$U = 0.9D \pm E \tag{21}$$

After the calculation of internal forces, maximum axial force (N_{max}) and maximum shear force (V_{nmax}) are checked for all columns to provide a ductile design. The equations of V_{nmax} and N_{max} are respectively given as Eq. (22) and (23).

$$V_{n\max} = \min \begin{Bmatrix} 5.5A \\ 0.2f_c^{'}A \end{Bmatrix} \tag{22}$$

$$N_{\max} = 0.5f_c^{'}A \tag{23}$$

The maximum axial force capacities of all beams are controlled according to Eq. (24).

$$N_{\max} = 0.1f_c^{'}A \tag{24}$$

If the mentioned ductile behaviour conditions are not provided, cross-sectional dimensions are reassigned, and all previous calculations are repeated.

After candidate cross-sections providing all previously presented design constraint are found, longitudinal reinforcements for tensile sections of beams are randomized. In that case, the exact value of effective depth of the beam can be calculated. If the maximum reinforcement ratio (ρ_{max}) given in equation (25) is exceeded, longitudinal reinforcements are randomized in compressive section, too. The maximum reinforcement area must be also less than 0.025.

$$\rho_{\max} = (0.75)(0.85)\beta_1 \frac{f_c^{'}}{f_y}\left(\frac{600}{600 + f_y}\right) \tag{25}$$

The minimum longitudinal reinforcement areas ($A_{s,min}$) calculated according to Eqs. (26) and (27) must be also considered as design constraints. These minimum reinforcement conditions must be checked for both tensile and compressive sections of the beam at supports and spans. Also, minimum reinforcements in compressive section at supports must be half of the area of reinforcement in tensile sections since dynamic internal forces may change their direction due to seismic loadings.

$$A_{s,\min} \geq \frac{\sqrt{f_c^{'}}}{4f_y} b_w d \tag{26}$$

$$A_{s,\min} \geq \frac{1.4}{f_y} b_w d \tag{27}$$

Positioning of the reinforcement bars is also checked according to maximum clear distance between reinforcements; a_ϕ given as Eq. (28). D_{max} and $\phi_{avarage}$ represent maximum aggregate diameter used in concrete and average diameter of longitudinal reinforcements. If rebars are not suitable in single line, placements of rebars in two lines are also possible. If positioning, maximum and minimum reinforcement constraints are not satisfied, the reinforcements are continued to randomize.

$$a_\varphi > \begin{cases} \varphi_{\text{avarage}} \\ 25\,mm \\ \frac{4}{3} D_{\max} \end{cases} \tag{28}$$

Randomly chosen reinforcements may not be optimum ones for the selected cross-section. By making several iterations, the best combination can be found. For a frame consisting of several structural members, finding the best combination of an assigned cross-section dimension may be too long and the chance to find it for all members is hard. For that reason, the reinforcement area found according to randomized rebar sizes ($A_{s\text{-random}}$) can be checked by comparing with the required reinforcement area ($A_{s\text{-needed}}$). This constraint is formulated as Eq. (29). Thus, all calculations including determination of static and dynamic loads are not needed to remake.

$$A_{s-needed} \leq A_{s-random} < (1+r)A_{s-needed} \tag{29}$$

r is a user defined value. If a small value of *r* is entered, the process may entrap to a range in which a physical solution cannot be found since the randomized area is provided via fixed rebar sizes. In that situation, the value of r can be increase after several iterations of this randomization stage. Also, if the solution cannot be found after maximum allowed iterations, the objective function defined as cost of RC frame is penalized by assigning a very big cost.

After the design of longitudinal reinforcements of the beam is finished, the stirrups are optimized. Nominal shear strength of concrete (V_c) and nominal shear strength of reinforcements (V_s) are presented as Eqs. (30) and (31), respectively. For ductility, V_s must not exceed Eq. (32). The optimized values for stirrups are size and the distance between bars (s). As design constraints, minimum shear reinforcement area ($(A_v)_{min}$) given as Eq. (33) and maximum spacing between shear reinforcements (s_{max}) given as Eqs. (34) and (35) must be provided.

$$V_c = \frac{\sqrt{f'_c}}{6} b_w d \tag{30}$$

$$V_s = \frac{A_v f_y d}{s} \tag{31}$$

$$V_s \leq 0.66\sqrt{f'_c} b_w d \tag{32}$$

$$\left(A_v\right)_{\min} = \frac{1}{3}\frac{b_w s}{f_y} \tag{33}$$

$$s_{\max} \leq \frac{d}{2} \tag{34}$$

$$s_{\max} \leq \frac{d}{4} \text{ if } V_s \geq 0.33\sqrt{f'_c} b_w d \tag{35}$$

After all design variables of beams are assigned with values providing all design constraints, the total cost of the beams is calculated. Then, the optimization process continues with the optimum design of reinforcements of columns.

The effect of slenderness is taken into consideration for slender columns by using the approximate design procedure presented in ACI 318. Since deflections need to represent the state just before the ultimate load in second order analysis, moment of inertias of members are reduced by 65% and 30% for beam and columns, respectively. In calculation of a moment magnification factor, the first factor is to find effective length factors in buckling (k). In order to obtain k, the value of Ψ presented as Eq. (36) at both ends of columns (Ψ_A at upper and Ψ_B at lower end) is firstly calculated. Then, the expressions given as Eqs. (37) – (39) are used for calculation of k for frames free to lateral movement.

$$\Psi_{A,B} = \frac{\sum \left(EI/l \right)_{column}}{\sum \left(EI/l \right)_{beam}} \tag{36}$$

$$\Psi_m = 0.5(\Psi_A + \Psi_B) \tag{37}$$

$$k = \frac{20 - \Psi_m}{20} \sqrt{1 + \Psi_m} \quad if\ \Psi_m < 2 \tag{38}$$

$$k = 0.9\sqrt{1 + \Psi_m} \quad if\ \Psi_m \geq 2 \tag{39}$$

The correction factor (C_m) used for correction of actually moment diagram to an equivalent moment diagram is calculated via Eq. (40). This factor must be bigger than 0.4. Also, it must be taken as 1.0 for the member with transverse loads between supports.

$$C_m = 0.6 + 0.4\frac{M_1}{M_2} \tag{40}$$

The critical buckling load is calculated via Eq. (41) by reducing the rigidity of column by 75%, and then, moment magnification factor (δ_s) used in multiplying the design flexural moment is calculated according to Eq. (42). P_u is the factored axial force.

$$P_c = \frac{\pi^2 EI}{(kl)^2} \tag{41}$$

$$\delta_s = \frac{C_m}{1 - \frac{P_u}{0.75P_c}} \tag{42}$$

For all columns which have a smaller flexural moment value than the minimum moment (M_{min}) formulated as Eq. (43), the minimum value calculated according to minimum eccentricity is used.

$$M_{\min} = P_u(15 + 0.03h) \text{ (h must be in mm)} \tag{43}$$

The positioning of rebars is also checked for columns. A symmetric design of rebars by assignment of longitudinal reinforcements randomly for both (upper and lower) end fibres. The positioning conditions are a little different than beams as seen in Eq. (44).

$$a_\varphi > \begin{cases} 1.5\varphi_{\text{avarage}} \\ 40\,mm \\ \frac{4}{3}D_{\max} \end{cases} \tag{44}$$

Additional web reinforcements are randomly assigned if the clear distance between upper and lower reinforcements is more than 150 mm. The randomizing process of rebars continues until placement of requirements, minimum (ρ_{min}=0. 01) and maximum (ρ_{max}=0. 06) reinforcement ratios are provided. If not, the process of assignment of rebars of columns iteratively repeated.

As presented in Chapter 4, distance from extreme compressive fibre to neutral axis (c) is iteratively scanned in the calculation of axial force and flexural moment capacities. The c value for the lowest axial force ensuring the required axial force value is stored. For that capacity, the flexural moment capacity ($M_{capacity}$) is defined. As done in the required reinforcement area of beams, the randomization of rebars of columns is repeated until Eq. (45) is provided. Mrequired is the needed minimum capacity calculated according to internal forces combination of static and dynamic analysis.

$$M_{required} \leq M_{capacity} < (1+r)M_{required} \tag{45}$$

Shear reinforcements of columns are also designed similarly to beams. Then, costs of columns are calculated. As the final calculation, total cost of the structure defined as the objective function (OF) of the problem is calculated.

Cost of an element (C_e) is calculated according to Eq. (46). In calculations, A_g represents area of cross-section, A_{st} represents area of non-prestressed longitudinal reinforcement, A_v represents area of shear reinforcement spacing s, u_{st} represents length of shear reinforcement spacing s and l_e represents the length of element. OF is the sum of all member costs as given in Eq. (47).

$$C_e = \left(A_g - A_{st}\right) l_e C_c + \left(A_{st} + \frac{A_v}{s} u_{st}\right) l_e \gamma_s C_s \tag{46}$$

$$OF = \sum_{i=1}^{n} \left(C_e\right)_i \text{ n:number of elements} \tag{47}$$

Since additional random search iterations are used in the presented method, the design variables stored in HM matrix are solutions which generally provide design constraints. For that reason, in the modified HS method with additional random search iterations, a new harmony vector is generated by using the same procedure after the generation of initial harmony memory matrix.

In the modified harmony search method, the boundary limits of solution range of cross-section dimensions are updated according to best existing harmony vector. In order to prevent trapping on local optimum, the initial boundary conditions are used for a structural member with 50% (similar factor as harmony memory considering rate) possibility. The lower or upper limit of the range is updated with equal possibility (similar factor as pitch adjusting rate) with the values of the best harmony vector. If the newly generated vector has a smaller objective function than existing worst one, it is replaced with it. Iterations continue until the number of generations reaches to a maximum iteration number. The all methodology is shown in the flowchart given as Fig. 3.

NUMERICAL EXAMPLES

The presented method was applied for two numerical examples; a two-span two-story symmetrical RC frame and a three-span three-story RC frame. Three different earthquake records were used, and the information about the records are given in Table 1. The value of design constants and the ranges used for both examples are given as Table 2. Discrete variables for dimensions and rebar sizes were used. The dimensions are assigned as multiples of 50 mm, and rebar sizes are taken as multiples of 2 mm. Each span is loaded with 15 kN/m dead load (D) and 5 kN/m live load (L). As the trapezium shape of distributed loads, the ratio of *a* and *l* is taken as ¼.

Numerical Example 1: Two Span-two Story Symmetric RC Frame

Two-span two-story RC frame seen in Fig. 4 has the optimum results given in Tables 3 and 4 for columns and beam members, respectively. For the optimum results of beams, LJ and RJ represent left and right joint of the element. The total cost of the RC frame is found as 304.78 unit money (Nigdeli & Bekdaş, 2016).

Numerical Example 2: Three Span-three Story RC Frame

Optimum results for a three-span three-story structure given as Fig. 5 is presented. The optimum results are shown as Table 5 and 6 for column and beam elements, respectively. The total cost of the optimum design is 724.54 unit money. The optimum value for the first five attempts (1148 unit money) is nearly 58% higher than the optimum cost. This means that a random design that can be also possible design of an engineer may be too big according to optimized results. This situation shows the importance of the optimization (Bekdaş & Nigdeli, 2017).

Figure 3. Flowchart of the presented methodology
(Bekdaş & Nigdeli, 2017)

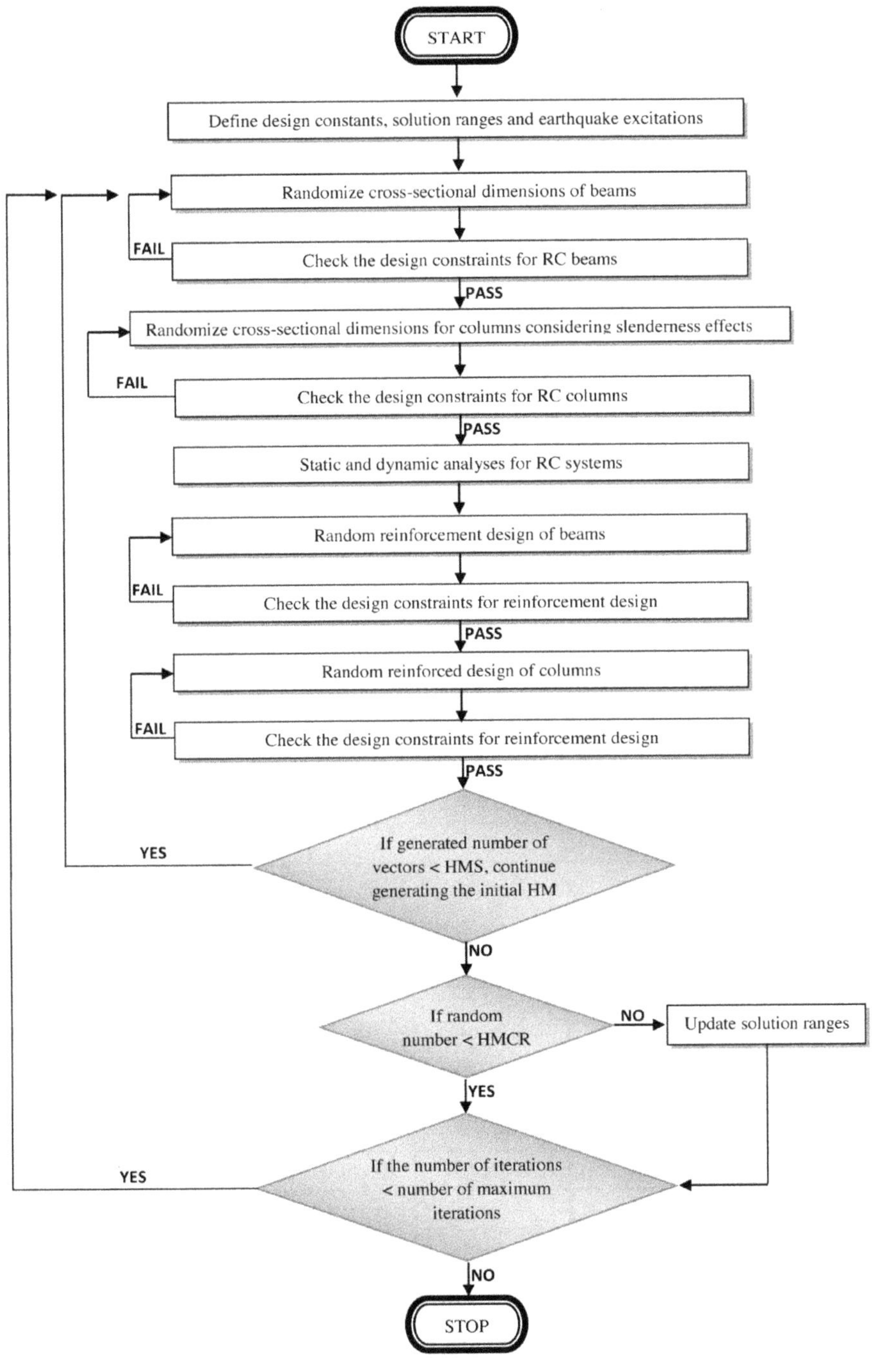

Table 1. Earthquake records

Earthquake	Date	Station	Component	PGA(g)
Imperial Valley	1940	117 El Centro	I-ELC180	0.313
Northridge	1994	24514 Sylmar	SYL360	0.843
Loma Prieta	1989	16 LGPC	LGP000	0.563

Note: Earthquake records were taken from PEER NGA DATABASE (http://peer.berkeley.edu/nga/)

Table 2. Design constants and ranges of design variables

Definition	Symbol	Unit	Value
Range of web width	b_w	mm	250-400
Range of height	h	mm	300-600
Clear cover	c_c	mm	30
Range of reinforcement	ϕ	mm	16-30
Range of shear reinforcement	ϕ_v	mm	8-14
Max. aggregate diameter	D_{max}	mm	16
Yield strength of steel	f_y	MPa	420
Comp. strength of concrete	f_c'	MPa	25
Elasticity modulus of steel	E_s	MPa	200000
Specific gravity of steel	γ_s	t/m^3	7.86
Specific gravity of concrete	γ_c	t/m^3	2.5
elastic response parameter	R	-	8.5
Cost of the concrete per m^3	C_c	unit	40
Cost of the steel per ton	C_s	unit	400

Table 3. Optimum design of columns (numerical example 1)

Element Number	b_w (mm)	h (mm)	Bars in Each Face	Shear Reinforcement Diameter/Distance (mm)
1	250	300	2Φ10+ 2Φ12	Φ8/120
2	250	300	2Φ10+ 2Φ12	Φ8/120
3	250	300	2Φ10+ 2Φ12	Φ8/120
6	250	300	2Φ10+ 2Φ12	Φ8/120
7	250	300	2Φ10+ 2Φ12	Φ8/120
8	250	300	2Φ10+ 2Φ12	Φ8/120

(Nigdeli & Bekdaş, 2016)

Figure 4. Model of first numerical example
(Nigdeli & Bekdaş, 2016)

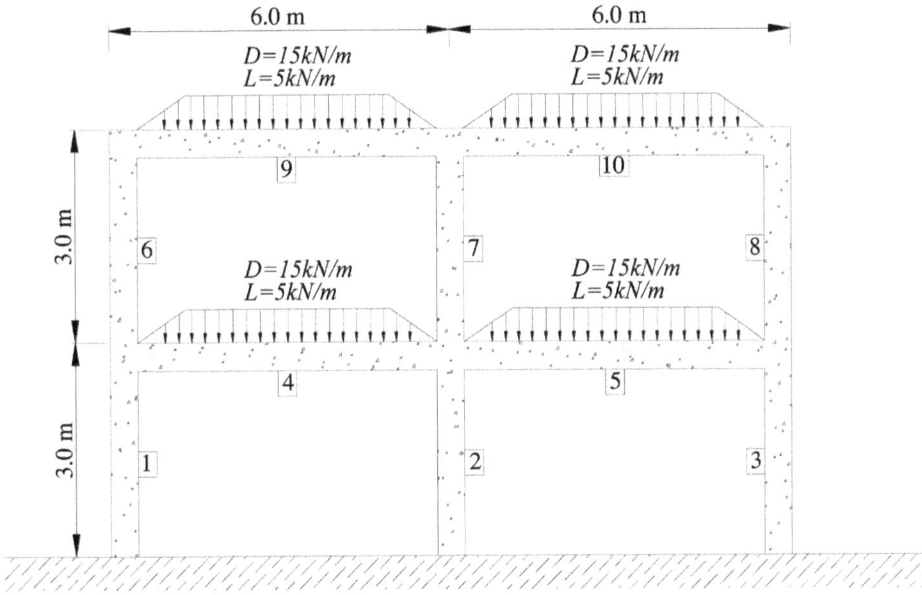

Table 4. Optimum design of beams (Numerical Example 1)

Element Number	b_w (mm)	h (mm)	Bars in Comp. Section	Bars in Tensile Section	Shear Reinforcement Diameter/Distance (mm)
LJ4			1Φ18+1Φ14 +1Φ16	1Φ20+1Φ30+1Φ14	
4	250	300	2Φ12	1Φ20+1Φ16	Φ8/120
RJ4-LJ5			2Φ16+1Φ12 +1Φ14	1Φ24+1Φ18+1Φ28	
5	250	300	2Φ12	1Φ20+1Φ16	Φ8/120
RJ5			1Φ18+1Φ14 +1Φ16	1Φ20+1Φ30+1Φ14	
LJ9			1Φ18+ 1Φ16	1Φ12+2Φ14+1Φ24	
9	250	300	2Φ12	1Φ14+1Φ26	Φ8/120
RJ9-LJ10			1Φ16+1Φ14 +1Φ20	1Φ22+1Φ24+1Φ16+1Φ18	
10	250	300	2Φ12	1Φ14+1Φ26	Φ8/120
LJ10			1Φ18+ 1Φ16	1Φ12+2Φ14+1Φ24	

(Nigdeli & Bekdaş, 2016)

Figure 5. Model of second numerical example
(Bekdaş & Nigdeli, 2017)

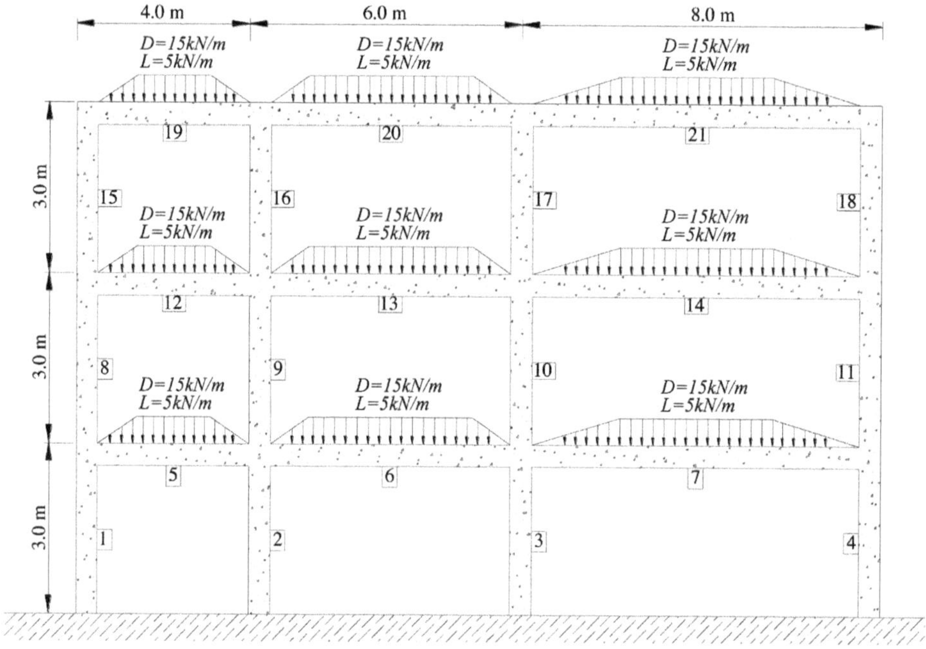

Table 5. Optimum design of columns (numerical example 2)

Element Number	b_w (mm)	h (mm)	Bars in Each Face	Shear Reinforcement Diameter/Distance (mm)
1	250	300	2Φ10+ 2Φ12	Φ8/120
2	250	350	3Φ10+ 1Φ12	Φ8/150
3	250	350	1Φ12+ 1Φ18+1Φ10	Φ8/150
4	250	300	2Φ10+ 2Φ12	Φ8/120
8	250	300	2Φ10+ 2Φ12	Φ8/120
9	250	300	2Φ10+ 2Φ12	Φ8/120
10	250	300	4Φ10+ 1Φ12	Φ8/120
11	250	300	2Φ10+ 2Φ12	Φ8/120
15	250	300	2Φ10+ 2Φ12	Φ8/120
16	250	300	2Φ10+ 2Φ12	Φ8/120
17	250	300	1Φ14+ 2Φ12	Φ8/120
18	250	300	3Φ18	Φ8/120

(Bekdaş & Nigdeli, 2017)

Table 6. Optimum design of beams (numerical example 2)

Element Number	b_w (mm)	h (mm)	Bars in Comp. Section	Bars in Tensile Section	Shear Reinforcement Diameter/Distance (mm)
LJ5			1Φ10+2Φ14 +1Φ16	1Φ24+1Φ28	
5	250	350	2Φ10+ 1Φ12	1Φ14+1Φ16+1Φ12	Φ8/150
RJ5-LJ6			1Φ10+1Φ26	1Φ16+ 2Φ14+1Φ28	
6	250	350	2Φ10+ 1Φ12	1Φ20+ 1Φ14	Φ8/150
RJ6-LJ7			1Φ22+1Φ14 +1Φ18	1Φ20+1Φ18+1Φ26+1Φ22	
7	250	450	1Φ10+1Φ12 +1Φ14	1Φ24+ 1Φ20	Φ8/200
RJ7			1Φ20+ 1Φ22	1Φ14+1Φ16+1Φ28+1Φ20	
LJ12			1Φ20+ 1Φ12	1Φ30+ 1Φ12	
12	250	450	1Φ16+ 1Φ14	1Φ12+ 1Φ22	Φ8/200
RJ12-LJ13			4Φ12	1Φ28+ 1Φ18	
13	250	400	4Φ10	1Φ14+ 1Φ12+1Φ16	Φ8/170
RJ13-LJ14			1Φ28+ 1Φ14	2Φ28+ 1Φ18	
14	250	400	1Φ16+ 1Φ12	1Φ10+ 2Φ12+1Φ26	Φ8/170
RJ14			1Φ16+1Φ22 +1Φ12	1Φ10+1Φ22+1Φ28+1Φ16	
LJ19			1Φ16+ 1Φ12	1Φ10+ 1Φ18+1Φ14	
19	250	400	2Φ14	2Φ14	Φ8/170
RJ19-LJ20			1Φ12+ 1Φ22	1Φ10+ 2Φ22+1Φ12	
20	250	300	2Φ12	1Φ10+1Φ16+1Φ12+1Φ14	Φ8/120
RJ20-LJ21			1Φ24+1Φ18	1Φ18+1Φ26+1Φ24+1Φ12	
21	250	400	1Φ12+ 1Φ16	1Φ24+ 1Φ26	Φ8/170
LJ21			1Φ20+ 1Φ12	1Φ26+ 1Φ20	

(Bekdaş & Nigdeli, 2017)

REFERENCES

ACI-318. (2005). Building code requirements for structural concrete and commentary, metric version, *American Concrete Institute*.

Bekdaş, G., & Nigdeli, S. M. (2017, February). Modified Harmony Search for Optimization of Reinforced Concrete Frames. In *International Conference on Harmony Search Algorithm* (pp. 213-221). Springer, Singapore. 10.1007/978-981-10-3728-3_21

Geem, Z. W., Kim, J. H., & Loganathan, G. V. (2001). A new heuristic optimization algorithm: Harmony search. *Simulation*, *76*(2), 60–68. doi:10.1177/003754970107600201

Mathworks. (2018). MATLAB R2018a. The MathWorks Inc., Natick, MA.

Nigdeli, S. M., & Bekdaş, G. (2016). Detailed Optimum Design of Reinforced Concrete Frame Structures. *WSEAS Transactions on Applied and Theoretical Mechanics*, *11*, 220–228.

PEER. (2005) Pacific earthquake engineering resource center: NGA database. University of California, Berkeley. Retrieved from http://peer.berkeley.edu/nga. Accessed November 2011.

Chapter 7
Optimum Design of Reinforced Concrete Retaining Walls:
Optimization of RC Retaining Walls

ABSTRACT

In this chapter, the optimization of reinforced concrete (RC) retaining walls is presented. RC retaining walls are one of the structural types that are constructed on land and used for retaining soil backfill. Because of this reason, both structural and geotechnical limits are in progress in the optimization process. Additionally, the stability conditions against pressure of soils are the key constraints in the optimum design of RC retaining walls. The presented methodology in this chapter considers both static and dynamic soil pressures resulting from earthquakes. A computer code employing teaching-learning-based optimization algorithm is also given.

INTRODUCTION

Retaining walls that are constructed for retaining soil backfill is one of the reinforced concrete (RC) structural types. These types of structures crucially need optimization because of existing of high number of design constraints including both structural and geotechnical limit states. On solving the optimum design variables of RC retaining walls, metaheuristic algorithms can be easily used.

DOI: 10.4018/978-1-7998-2664-4.ch007

The optimization of RC retaining walls dates back to very early times as 80s (Rhomberg & Street, 1981; Alshawi, Mohammed, & Farid, 1988; Keskar & Adidam, 1989; Dembicki & Chi, 1989; Pochtman, Zhmuro, & Landa, 1988). Then, metaheuristic optimization techniques are employed in 2000s (Chau & Albermani 2003; Ceranic, Fryer, & Baines, 2001; Yepes, Alcala, Perea, & González-Vidosa, 2008; Camp & Akin 2011; Temur & Bekdaş 2016; Kayabekir, Bekdaş, Nigdeli, & Temür, 2016; Bekdaş, Nigdeli, Temür, & Kayabekir, 2016).

In this chapter, the methodology developed by Bekdaş et al. (2016) is presented, and a computer code employing teaching-learning-based optimization (TLBO) (Rao, Savsani, & Vakharia, 2011) is given as an appendix.

OPTIMUM DESIGN OF RC RETAINING WALLS

A soil retaining wall is given as Fig. 1 with the soil pressure resulting from active earth pressure (P_A), passive earth pressure (P_P), bearing stress (P_T), active earth pressure (P_{AD}), active earth pressure resulting from surcharge loads for static (Q_A) and dynamic load case (Q_{AD}). The other symbols in Fig.1 represent the following parameters:

W_v: The weight of the retaining wall.
W_T: The weight of backfill on the heel.
q: The surcharge loads.
q_{min}: The minimum stress under the footing.
q_{max}: The maximum stress under the footing.
H: The height of the wall.
γ_R: Specific gravity of soil.
ϕ_R: Internal friction angle.
c_R: Cohesion of retaining soil.

The optimization of RC retaining walls involves two limit states. In addition to structural limit states considered in all optimization approaches of RC members, the geotechnical state limits must also be taken as design constraints. These constraints are related about providing the stability measures of retaining walls such as overturning, sliding and bearing capacity control.

The safety factor for overturning (SF_o) is computed according to Eq. (1).

Figure 1. A cantilever retaining wall and loading conditions
(Bekdaş et al., 2016)

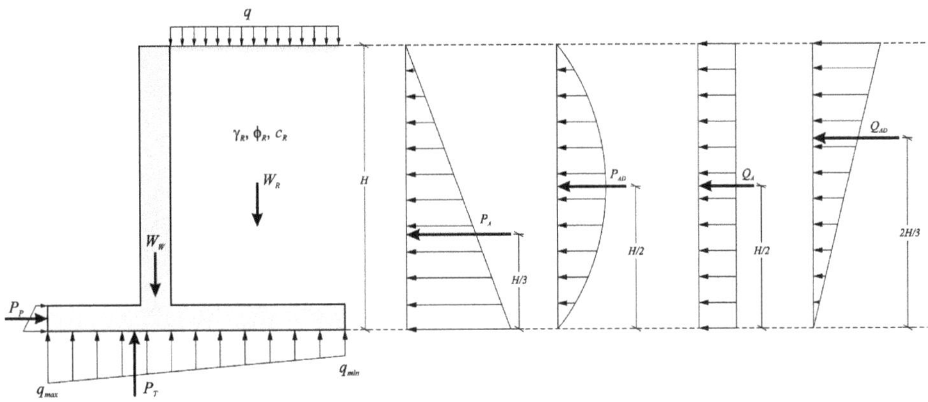

$$SF_O = \frac{\sum M_R}{\sum M_o} \quad (1)$$

$\sum M_R$ defines the sum of total moments that resist for overturning, and $\sum M_O$ is the sum of the moments resulting overturning. For safety, the passive pressure is not taken into consideration since these loads may be removed in time.

$$SF_O = \frac{\sum M_R}{\sum M_o} \quad (1)$$

The coefficient; k_a used in obtaining active earth pressure is defined according to the Rankine theory, and it is formulated as Eq. (2). β and θ represent the slope and internal friction angle of backfill, respectively.

$$k_a = \cos\beta \frac{\cos\beta - \sqrt{\cos^2\beta - \cos^2\theta}}{\cos\beta + \sqrt{\cos^2\beta - \cos^2\theta}} \quad (2)$$

The second stability condition is related to providing a safety factor for sliding failure (SF_s) as formulated in Eq. (3), where the sum of resisting forces ($\sum F_R$) and the sliding forces ($\sum F_D$) are given as Eqs. (4) and (5), respectively.

$$SF_S = \frac{\sum F_R}{\sum F_D} \tag{3}$$

$$\sum F_R = \left(\sum W_W\right)\tan\left(\frac{2\varphi_{base}}{3}\right) + \frac{2Bc_{base}}{3} + P_P \tag{4}$$

$$\sum F_D = P_a \cos\beta \tag{5}$$

In the calculation of forces, ϕ_{base}, B and c_{base} are internal friction angle of the base soil, length of the base slab and adhesion of the base soil, respectively.

The passive earth pressure is calculated as Eq. (6), where γ_{base} and D_1 are specific gravity and depth of the base soil, respectively.

$$P_P = \frac{1}{2}\gamma_{base} D_1^2 k_p + 2c_{base} D_1 \sqrt{k_p} \tag{6}$$

As the final stability condition, the safety factor of the retaining wall for bearing failure (SF_B) is given as Eq. (7), where q_u is the bearing capacity of the soil checked according to maximum soil pressure (q_{max}). The extremum (maximum; q_{max} and minimum; q_{min}) soil pressures are found according to Eq. (8) according to eccentricity (e) of the moments defined as Eq. (9).

$$SF_B = \frac{q_u}{q_{\max}} \tag{7}$$

$$q_{\min,\max} = \frac{\sum V}{B}\left(1 \mp \frac{6e}{B}\right) \tag{8}$$

$$e = \frac{B}{2} - \frac{\sum M_R - \sum M_O}{\sum V} \tag{9}$$

$\sum V$ is the sum of all vertical loads. The engineering problem includes eight design variables. Four of these variables are related to the geometry of the wall (dimensions as seen in Fig. 2), and the remaining ones are the properties about the reinforcement bars (rebars) of the wall. All variables are also shown in Table 1.

The design constraints are presented in Table 2. The first four constraints are related to the geotechnical state limits, and other ones are defined for controlling structural state limits. In Table 2, M_u, V_u, A_s and S define flexural moment in critical sections, shear force in critical section, the reinforcement area in critical sections and spacing of rebars, respectively.

The objective function is defined as the total material cost of the RC retaining wall is given as Eq. (10). In this equation, C_c is the cost of the concrete per m^3 including transportation and forming expenses. C_s is the cost of the steel per m^3 including transportation and labor work expenses. V_c represents the volume of the concrete and W_s denotes the total weight of the reinforcements.

$$\min f(X) = C_c V_c + C_s W_s \tag{10}$$

Figure 2. Geometrical design variables of a retaining wall
(Bekdaş et al., 2016)

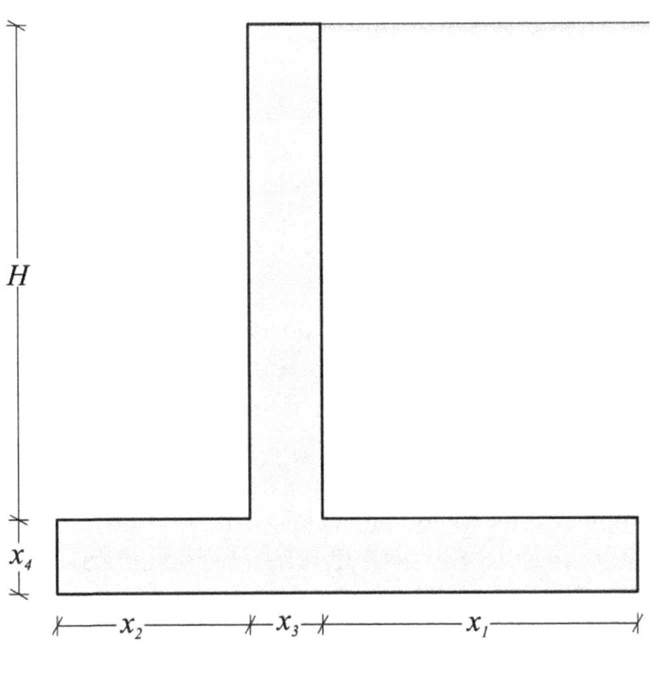

Table 1. The definition of design variables

	Definition	Design Variables
Geometrical design variables	The length of the heel	X_1
	The length of the toe	X_2
	The thickness of the stem	X_3
	The height of the footing	X_4
Design variables about rebars	The size of reinforcements of stem	X_5
	The spacing of the bars in stem	X_6
	The size of reinforcements of footing	X_7
	The spacing of the bars of footing	X_8

Table 2. The definition of design constraints

Definition	Constraints
Safety for overturning	$g_1(X)$: $SF_{O,design} \geq SF_O$
Safety for sliding	$g_2(X)$: $SF_{S,design} \geq SF_S$
Safety for maximum bearing capacity	$g_3(X)$: $SF_{B,design} \geq SF_B$
Safety for maximum bearing capacity, q_{min}	$g_4(X)$: $q_{min} \geq 0$
Flexural moment capacity in critical sections, M_d	$g_{5\text{-}6}(X)$: $M_d \geq M_u$
Shear force capacity in critical sections, V_d	$g_{7\text{-}8}(X)$: $V_d \geq V_u$
Minimum reinforcement area in critical sections, A_{smin}	$g_{9\text{-}10}(X)$: $A_s \geq A_{smin}$
Maximum reinforcement area in critical sections, A_{smax}	$g_{11\text{-}12}(X)$: $A_s \leq A_{smax}$
Maximum spacing of bars in critical sections, S_{max}	$g_{13\text{-}14}(X)$: $S \leq S_{max}$
Minimum spacing of bars in critical sections, S_{min}	$g_{15\text{-}16}(X)$: $S \geq S_{min}$
Minimum clear cover, c_c	$g_{17}(X)$: $c_c \geq 40\ mm$

The Numerical Examples

As numerical examples, three cases of design variable ranges are presented. The design constants and ranges are given in Table 3. In the second case, the maximum limits are lower than the first case for the dimensions of heel and toe of the wall. In the last case, toe projection has a more limited range to idealize the restriction of place of construction.

The optimum results for geometrical variables and reinforcements are presented in Tables 4 and 5, respectively. The rebar optimization of the retaining wall is given for 3 sections of the stem (by dividing it to equal lengths) and

Table 3. The design constants and design variables

Definition	Symbol	Unit	Value
Height of stem	H	m	5.6
Yield strength of steel	f_y	MPa	420
Compressive strength of concrete	f'_c	MPa	25
Concrete cover	c_c	mm	60
Max. aggregate diameter	D_{max}	mm	16
Elasticity modulus of steel	E_s	GPa	200
Specific gravity of steel	γ_s	t/m^3	7.85
Specific gravity of concrete	γ_c	kN/m^3	23.5
Cost of concrete per m^3	C_c	unit	104
Cost of steel per ton	C_s	unit	1751
Surcharge load	q	kPa	10
Backfill slope angle	β	°	0
Internal friction angle of retained soil	ϕ_R	°	30
Internal friction angle of base soil	ϕ_B	°	0
Unit weight of retained soil	γ_R	kN/m^3	18
Unit weight of base soil	γ_B	kN/m^3	18
Cohesion of retained soil	c_R	kPa	0
Cohesion of base soil	c_B	kPa	0
Depth of the soil in front of wall	D	m	0
Bearing capacity of the soil	q_U	kPa	250
Safety for overturning stability	$SF_{O,design}$	-	2
Safety for sliding stability	$SF_{S,design}$	-	2
Safety for overturning stability under earthquake	$SF_{O,design}$	-	1.3
Safety for sliding stability under earthquake	$SF_{S,design}$	-	1.3
Safety for bearing capacity	$SF_{B,design}$	-	1.0
Range of stem thickness	X_3	m	0.2-1
Range of heel and toe projection for case 1	X_1-X_2	m	0.2-4
Range of heel and toe projection for case 2	X_1-X_2	m	0.2-3
Range of heel projection for case 3	X_1	m	0.2-3
Range of toe projection for case 3	X_2	m	0.2-1
Range of footing thickness	X_4	m	0.2-1
Range of diameter of reinforcing bars of stem	X_5	mm	16-50
Range of diameter of reinforcing bars of footing	X_7	mm	16-50

(Bekdaş et al., 2016)

Table 4. Optimum geometrical design variables and objective function

Design Variables	Optimum Values of Design Variables (m)		
	Case 1	Case 2	Case 3
X_1	3.30	3.00	3.00
X_2	0.70	1.60	1.00
X_3	0.40	0.55	0.60
X_4	0.40	0.40	0.40
Optimum cost (unit)	731.09	818.23	830.11

(Bekdaş et al., 2016)

Table 5. Optimum reinforcements

	Case 1	Case 1	Case 1
	Bar Size(*mm*) / Spacing(*mm*)	Bar Size(*mm*) / Spacing(*mm*)	Bar Size(*mm*) / Spacing(*mm*)
$X_{5(1)}/X_{6(1)}$	ϕ12/160	ϕ16/200	ϕ14/140
$X_{5(2)}/X_{6(2)}$	ϕ14/150	ϕ16/200	ϕ14/140
$X_{5(3)}/X_{6(3)}$	ϕ28/180	ϕ14/70	Φ22/190
$X_{7(1)}/X_{8(1)}$	ϕ16/180	ϕ12/160	ϕ12/120
$X_{7(2)}/X_{8(2)}$	ϕ30/190	ϕ26/190	ϕ26/160

(Bekdaş et al., 2016)

2 critical sections of the footing. The optimum cost values are also shown in Table 4. The optimum costs are for unit length of the wall.

In the first case, the maximum bound of the heel of the retaining wall is defined as 4 m, and the optimum cost is 731.09 units while the optimum value of heel is found as 3.3 m. The maximum limit of the heel is reduced below the optimum value of Case 1 for the second case. Due to that, the optimum length of the toe increases to 1.6 m from 0.7 m, and the optimum cost increases by 12%. Because of the limitation of the heel, the only option in the optimum design methodology is to increase the length of toe in order to provide the required bearing capacity.

In the last case, toe value is also limited to 1 m in additional to the limitation of the heel. This time, the thicknesses of the stem and footing show an increasing manner, and the optimum cost increase is seen.

The Matlab (2018) code employing TLBO is presented in Appendix 1 for a wall without toe. The code includes a sub-code called "calculations_retaining_wall" and it is given as Appendix 2.

REFERENCES

Alshawi, F. A. N., Mohammed, A. I., & Farid, B. J. (1988). Optimum design of tied-back retaining walls. *Struct. Eng*, *66*(6), 97–105.

Bekdaş, G., Nigdeli, S. M., Temur, R., & Kayabekir, A. E. (2016). Restricted Optimum Design of Reinforced Concrete Retaining Walls. *International Journal of Theoretical and Applied Mechanics*, *1*, 326–332.

Camp, C. V., & Akin, A. (2011). Design of retaining walls using big bang–big crunch optimization. *Journal of Structural Engineering*, *138*(3), 438–448. doi:10.1061/(ASCE)ST.1943-541X.0000461

Ceranic, B., Fryer, C., & Baines, R. W. (2001). An application of simulated annealing to the optimum design of reinforced concrete retaining structures. *Computers & Structures*, *79*(17), 1569–1581. doi:10.1016/S0045-7949(01)00037-2

Chau, K. W., & Albermani, F. (2003). Knowledge-based system on optimum design of liquid retaining structures with genetic algorithms. *Journal of Structural Engineering*, *129*(10), 1312–1321. doi:10.1061/(ASCE)0733-9445(2003)129:10(1312)

Dembicki, E., & Chi, T. (1989). System analysis in calculation of cantilever retaining walls. *International Journal for Numerical and Analytical Methods in Geomechanics*, *13*(6), 599–610. doi:10.1002/nag.1610130603

Kayabekir, A. E., Bekdaş, G., Nigdeli, S. M., & Temür, R. (2016, September). Optimum design of retaining walls under static and dynamic loads, In *International Science Symposium (ISS2016)*.

Keskar, A. V., & Adidam, S. R. (1989). Minimum cost design of a cantilever retaining wall. *Indian Concrete Journal*, *63*(8).

Mathworks. (2018). MATLAB R2018a. The MathWorks Inc., Natick, MA.

Pochtman, Y. M., Zhmuro, O. V., & Landa, M. S. (1988). Design of an optimal retaining wall with anchorage. *Soil Mechanics and Foundation Engineering*, *25*(5), 508–510. doi:10.1007/BF01786646

Rao, R. V., Savsani, V. J., & Vakharia, D. P. (2011). Teaching–learning-based optimization: A novel method for constrained mechanical design optimization problems. *Computer Aided Design*, *43*(3), 303–315. doi:10.1016/j.cad.2010.12.015

Rhomberg, E. J., & Street, W. M. (1981). Optimal design of retaining walls. *Journal of the Structural Division*, *107*(5), 992–1002.

Temur, R., & Bekdaş, G. (2016). Teaching learning-based optimization for design of cantilever retaining walls. *Structural Engineering and Mechanics*, *57*(4), 763–783. doi:10.12989em.2016.57.4.763

Yepes, V., Alcala, J., Perea, C., & González-Vidosa, F. (2008). A parametric study of optimum earth-retaining walls by simulated annealing. *Engineering Structures*, *30*(3), 821–830. doi:10.1016/j.engstruct.2007.05.023

APPENDIX 1

```
clear all
clc
format compact
format long
% 1.ENTER DATA OF PROBLEM
gama_soil=18;     % Specific gravity of soil, [kN/m^3]
gama_conc=25;     % Specific gravity of concrete, [kN/m^3]
gama_steel=7.85; % Unit volume weight of stem, [t/m^3]
H=5.6;            % Height of the wall [m]
b=1;              % Unit length [m]
d_prime=0.06;     % Distance from extreme compression fiber to
centroid of longitudinal compression reinforcement
phi=30;           % Internal friction angle of retained soil
q=10;             % Surcharge load
fctd=1.15;        % MPa
fyd=360;          % MPa
fcd=16.7;         % MPa
romin=0.002;      % Lower limit of the reinforcement ratio
k_fric=0.6;       % Friction coefficient of base soil
qs=250;           % Safety for bearing capacity
NSb=2;            % The number of points of reinforced concrete
design along the base.
NSw=3;            % The number of points of reinforced concrete
design along the wall height.
Cs=1300;          % Unit cost of reinforcement [unit/ton]
Cc=120;           % Unit cost of concrete [unit/m^3]
% Required parameters for earthquake load calculation
Z=3;              % Ground group
A0=0.3;           % Effective Ground Acceleration Coefficient
I=1;              % Structure Importance Coefficient
% User defined algorithm parameters
pn=20;          % population number
maxiter=20000; % maximum iteration number
% Ultimate limits of design variables
hwmin=0.2; hwmax=1; % lower and upper limits of wall thickness
hbmin=0.2; hbmax=2; % lower and upper limits of base thickness
B1min=0;   B1max=0;   % lower and upper limits of toe
B2min=0.2;   B2max=4;    % lower and upper limits of heel
% 1.1. PRECALCULATIONS NEEDED for OPTIMIZATION PROCESS
% Active earth pressure coefficient according to Rankine
theory:
Ka=(cosd(phi))^2/(1+sind(phi))^2;
% Active earth pressure coefficient including eartquake loads:
Ch=0.2*(I+1)*A0; % Equivalent Earthquake Coefficient in
horizontal direction
```

```
Cv=2*Ch/3;        % Equivalent Earthquake Coefficient in
vertical direction
lamda1=atand(Ch/(1-Cv));
lamda2=atand(Ch/(1+Cv));
K_at1=((1+Cv)*(cosd(phi-lamda2))^2)/((cosd(lamda2))^2*(1+sqrt(s
ind(phi)*(sind(phi)-tand(lamda2)*cosd(phi))))^2);
K_at2=((1-Cv)*(cosd(phi-lamda1))^2)/((cosd(lamda1))^2*(1+sqrt(s
ind(phi)*(sind(phi)-tand(lamda1)*cosd(phi))))^2);
if K_at1>K_at2
    K_at=K_at1;
else
    K_at=K_at2;
end
K_ad=K_at-Ka;     % Dynamic active coefficient
% Dynamic force considering surcharge load:
Q_AD=K_ad*q*H;
% Variation of dynamic pressure along wall height
syms z
P_ad=3*K_ad*(1-z/H)*gama_soil*z;
% Dynamic force considering soil pressure:
P_AD=double(int(P_ad,z,0,H));
%==============================================================
===========
% 2. GENERATING OF INITIAL SOLUTION MATRIX
for I=1:pn
    % 2.1. DEFINITION OF DESING VARIABLES RANDOMLY
    hw=(round((hwmin+rand*(hwmax-hwmin))*20))/20;
    hb=(round((hbmin+rand*(hbmax-hbmin))*20))/20;
    B1=(round((B1min+rand*(B1max-B1min))*20))/20;
    B2=(round((B2min+rand*(B2max-B2min))*20))/20;
    % 2.2. OTHER CALCULATIONS
    for i=1:NSw+NSb
        if i<=NSw
            hf(i)=hw; % height of wall cross-section
        else
            hf(i)=hb; % height of base cross-section
        end
    end
    d=hf-d_prime*ones(1,NSw+NSb); % effective depth vector
    P_A=0.5*Ka*gama_soil*(H+hb)^2; % back-wall force resulting
from soil load
    Q_A=Ka*q*(H+hb);                % back-wall force resulting
from surcharge load
    W1=H*B2*gama_soil;             % weight of soil
    W2=H*hw*gama_conc;             % weight of wall
    W3=(B2+hw+B1)*hb*gama_conc;    % weight of base
    % 2.2.1. Calculation of soil stresses (qsmax-qsmin)
    % In the static case
    N=W1+W2+W3+q*(B2+hw);
```

```
M=P_A*((H+hb)/3)+0.5*Q_A*(H+hb)-W1*((B2+hw+B1)*0.5-
0.5*B2)+W2*0.5*(B2+hw);
    qsmax=N/(B2+hw+B1)+6*M/(B2+hw+B1)^2;
    qsmin=N/(B2+hw+B1)-6*M/(B2+hw+B1)^2;
    % In the dynamic case
    Me=P_AD*((H/2)+hb)+Q_AD*(H*(2/3)+hb);
    qs_add1=Me*6/(B2+hw+B1)^2;
    qs_add2=-Me*6/(B2+hw+B1)^2;
    qsmax_e=qsmax+qs_add1;
    qsmin_e=qsmin+qs_add2;
    % 2.2.2. Calculation of safety factors
    % In the static case
    % The safety factor for overturning (SF_O)
    MD=P_A*(H+hb)/3+Q_A*(H+hb)/2;                    % the
overturning moments
    MR=W1*(B2/2+hw+B1)+W2*(hw/2+B1)+W3*(B2+hw+B1)/2; % the
resisting forces
    SF_O=MR/MD;
    % The safety factor for sliding failure (FS)
    FD=P_A+Q_A;             % the sliding forces
    FR=k_fric*(W1+W2+W3); % the resisting forces
    SF_S=FR/FD;
    % In the dynamic case considering eartquake forces
    % The safety factor for overturning (SFe_O)
    MeD=P_AD*(H/2+hb)+Q_AD*(H*2/3+hb)+MD;
    MeR=MR;
    SFe_O=MeR/MeD;
    % The safety factor for overturning (SFe_S)
    FeD=P_AD+Q_AD+FD;
    FeR=FR;
    SFe_S=FeR/FeD;
    % 2.2.3. Calculation of cross-sectional effects
    % Cross-sectional effects on the wall due to soil load:
    for i=1:NSw
        % Axial forces for every part of the wall
        Nw(i,1)=hw*(i*H/NSw)*gama_conc;
        % Shear forces for every part of the wall
        Vw(i,1)=Ka*gama_soil*(i*H/NSw)^2*0.5+Ka*q*(i*H/NSw);
        % Bending moments for every part of the wall
        Mw(i,1)=Ka*gama_soil*(i*H/NSw)^2*0.5*(i*H/
NSw^2)+Ka*q*(i*H/NSw)*(i*H/(2*NSw));
    end
    % Cross-sectional effects on the base due to soil load:
    L=(B2+hw+B1)*qsmin/(qsmax-qsmin); % parameter to be used in
proportion to find the base stress.
    for l=1:NSb
        % Shear forces for every part of the base
        Vb(l,1)=(q+H*gama_soil+hb*gama_conc)*(l*B2/NSb)-
```

```
qsmin*(l*B2/NSb)*0.5-(qsmin*(L+l*B2/NSb)/L)*(l*B2/NSb)*0.5;
        % Bending moments for every part of the base
        Mb(l,1)=((q+H*gama_soil+hb*gama_conc)*(l*B2/NSb)^2)/2-
(qsmin*(l*B2/NSb)^2)/3-((qsmin*(L+l*B2/NSb)/L)*(l*B2/NSb)^2)/6;
    end
    % Additional effects due to eartquake forces on the wall:
    for i=1:NSw
        Vwe(i,1)=double(int(P_ad,z,0,i*H/NSw))+K_ad*q*(i*H/
NSw)+(K_ad*q*i/NSw)*(H-i*H/NSw);
        x=double(int(P_ad*z,z,0,i*H/NSw)/int(P_ad,z,0,i*H/
NSw));
        Mwe(i,1)=double(int(P_ad,z,0,i*H/NSw))*((i*H/NSw)-x)+K_
ad*q*(i*H/NSw)*(2*H*i/(NSw*3))+(K_ad*q*i/NSw)*(H-(i*H/
NSw))*(H*i/(NSw*3));
    end
    % Additional effects due to eartquake forces on the base:
    L2=B2+hw+B1;
    for l=1:NSb
        if B2*l/NSb<L2/2
            Vbe(l,1)=(1-(2*B2*l/NSb)/L2)*(-qs_add2)*(B2*l/
NSb)*0.5+(-qs_add2)*(B2*l/NSb)*0.5;
            Mbe(l,1)=(1-(2*B2*l/NSb)/L2)*(-qs_add2)*(B2*l/
NSb)^2/6+(-qs_add2)*(B2*l/NSb)^2/3;
        else
            Vbe(l,1)=(-qs_add2)*L2/2*0.5-(qs_add1*(B2*l/
NSb-L2/2)/(L2/2))*(B2*l/NSb-L2/2)*0.5;
            Mbe(l,1)=(-qs_add2)*L2/2*0.5*(l*B2/NSb-
L2/2+2*L2/6)-(qs_add1*(B2*l/NSb-L2/2)/(L2/2))*(B2*l/
NSb-L2/2)^2/6;
        end
    end
    % 2.2.4. RC design of the cantilever retaining wall
    % Determination of bearing design moments
    % NOTE: Cross-sectional effects resulting from earthquake
effects shall be reduced by
    %       Ra = 1.5 coefficient in cross-sectional
calculation.
    ME=[Mwe;Mbe]/1.5;
    MH=[Mw;Mb];
    MU1=ME+MH;   % Moment combination 1
    MU2=1.6.*MH; % Moment combination 2
    NG=Nw;
    for r=1:NSw+NSb
        Md(r,1)=max(MU2(r,1), MH(r,1)+ME(r,1));
        % Depth of equivalent rectangular stress blocks
        a=d(r)*1000-sqrt((d(r)*1000)^2-2.*Md(r,1)*10^3/
(0.85*fcd*b));
        % Required reinforcement area [mm^2]
        As(r)=(0.85*fcd*b*1000*a/fyd);
```

```
    end
    smax=(1.5).*hf*1000; % upper limit vector of rebar spacing
[m]
    for i=1:NSw+NSb
        % Control of the minimum required reinforcement areas
        if As(i)<romin*(b*1000)*d(i)*1000
            As(i)=romin*(b*1000)*d(i)*1000;
        end
        for j=1:10
            % Generation of diameter of rebars
            R=10+j*2;
            % Generation of diameter of rebar spacings
            s=10*fix((pi*R^2/4*(1000/As(i)))/10);
            % Control of the rebar spacings considering
ultimate limits
            if s>smax(i) || s>200
                s=min(smax(i),200);
            end
            % Store these values to matrix
            D1(j,1)=R;
            D1(j,2)=s;
            D1(j,3)=(pi*R^2/4*(1000/s));
        end
        [k,t]=min(D1(:,3));
        D(i,:)=D1(t,:);
    end
    % Determination of shear design forces
    % NOTE: Cross-sectional effects resulting from earthquake
effects shall be reduced by
    %       Ra = 1.5 coefficient in cross-sectional
calculation.
    VE=[Vwe;Vbe]/1.5;
    VH=[Vw;Vb];
    VU1=VE+VH;   % Shear force combination1
    VU2=1.6.*VH; % Shear force combination2
    for r=1:NSw+NSb
        Vd(r,1)=max(VU2(r,1),VH(r,1)+VE(r,1)); % Shear design
force
    end
    Vcr=0.65*(fctd*1000)*b*d;  % Shear foce capacity of
concrete
    % 2.3 OBJECTIVE FUNCTION
    % Cost of concrete:
    COSTC=(hw*H+hb*(B1+B2+hw))*b*Cc;
    % Calculations for cost of reinforcing bars:
    COSTS=0;
    for i=1:NSw
        COSTS=COSTS+D(i,3)*(10^-6)*(H/NSw)*gama_steel*Cs;
    end
```

```
    for j=NSw+1:NSw+NSb
        COSTS=COSTS+D(j,3)*(10^-6)*(B2/NSb)*gama_steel*Cs;
    end
    Fx=COSTC+COSTS; % Objective function: Total cost for one
meter.
    % 2.4. PART OF PENALIZATION
    for i=1:NSw+NSb
        if Vcr<Vd(i,1)
            Fx=10^12;
        end
    end
    if qsmin<0 || qsmax>qs || SF_O<2 || SF_S<2 || SFe_O<1.3 ||
SFe_S<1.3
        Fx=10^12;
    end
    if Z~=4
        qs_e=1.5*qs;
        if qsmax_e>=qs_e
            Fx=10^12;
        end
    else
        if  qsmax_e>=qs
            Fx=10^12;
        end
    end
    if imag(Fx)>0 || imag(Fx)<0
        Fx=10^12;
    end
    % 2.5. INITIAL SOLUTION MATRIX
    OPT(I,1)=hw;
    OPT(I,2)=hb;
    OPT(I,3)=B1;
    OPT(I,4)=B2;
    OPT(I,5)=Fx;
end
for iter=1:maxiter
    %%%%%%%%%%%%%%%%%%%%%%%%%%%%%%%%%%%%%%%%Teacher pha
se%%%%%%%%%%%%%%%%%%%%%%%%%%%%%%%%%%%%%%%%
    for I=1:pn
        [pe,re]=min(OPT(5,:));
        best(1,:)=OPT(re,:);
        hw=OPT(I,1)+(round((rand*(best(1,1)-round(1+rand)*mean(
OPT(:,1))))*20))/20;
        hb=OPT(I,2)+(round((rand*(best(1,2)-round(1+rand)*mean(
OPT(:,2))))*20))/20;
        B1=OPT(I,3)+(round((rand*(best(1,3)-round(1+rand)*mean(
OPT(:,3))))*20))/20;
        B2=OPT(I,4)+(round((rand*(best(1,4)-round(1+rand)*mean(
OPT(:,4))))*20))/20;
```

```
        calculations_retaining_wall
    end
    %%%%%%%%%%%%%%%%%%%%%%%%%%%%%%%%%%%%%%Learner pha
se%%%%%%%%%%%%%%%%%%%%%%%%%%%%%%%%%%%%%%
    for I=1:pn
        v1=ceil(rand*pn);
        v2=ceil(rand*pn);
        hw=OPT(I,1)+(round(rand*(OPT(v1,1)-
(OPT(v2,1))))*20)/20;
        hb=OPT(I,2)+(round(rand*(OPT(v1,2)-
(OPT(v2,2))))*20)/20;
        B1=OPT(I,3)+(round(rand*(OPT(v1,3)-
(OPT(v2,3))))*20)/20;
        B2=OPT(I,4)+(round(rand*(OPT(v1,4)-
(OPT(v2,4))))*20)/20;
        calculations_retaining_wall
    end
end
```

APPENDIX 2

```
% 1. CONTROL OF ULTIMATE LIMITS
if hw>hwmax
    hw=hwmax;
end
if hw<hwmin
    hw=hwmin;
end
if hb>hbmax
    hb=hbmax;
end
if hb<hbmin
    hb=hbmin;
end
if B1>B1max
    B1=B1max;
end
if B1<B1min
    B1=B1min;
end
if B2>B2max
    B2=B2max;
```

```
end
if B2<B2min
    B2=B2min;
end
% 2. OTHER CALCULATIONS
for i=1:NSw+NSb
    if i<=NSw
        hf(i)=hw; % height of wall cross-section
    else
        hf(i)=hb; % height of base cross-section
    end
end
d=hf-d_prime*ones(1,NSw+NSb); % effective depth vector
P_A=0.5*Ka*gama_soil*(H+hb)^2; % back-wall force resulting from
soil load
Q_A=Ka*q*(H+hb);                % back-wall force resulting from
surcharge load
W1=H*B2*gama_soil;             % weight of soil
W2=H*hw*gama_conc;             % weight of wall
W3=(B2+hw+B1)*hb*gama_conc;    % weight of base
% 2.1. Calculation of soil stresses (qsmax-qsmin)
% In the static case
N=W1+W2+W3+q*(B2+hw);
M=P_A*((H+hb)/3)+0.5*Q_A*(H+hb)-W1*((B2+hw+B1)*0.5-
0.5*B2)+W2*0.5*(B2+hw);
qsmax=N/(B2+hw+B1)+6*M/(B2+hw+B1)^2;
qsmin=N/(B2+hw+B1)-6*M/(B2+hw+B1)^2;
% In the dynamic case
Me=P_AD*((H/2)+hb)+Q_AD*(H*(2/3)+hb);
qs_add1=Me*6/(B2+hw+B1)^2;
qs_add2=-Me*6/(B2+hw+B1)^2;
qsmax_e=qsmax+qs_add1;
qsmin_e=qsmin+qs_add2;
% 2.2. Calculation of safety factors
% In the static case
% The safety factor for overturning (SF_O)
MD=P_A*(H+hb)/3+Q_A*(H+hb)/2;                          % the
overturning moments
MR=W1*(B2/2+hw+B1)+W2*(hw/2+B1)+W3*(B2+hw+B1)/2; % the
resisting forces
SF_O=MR/MD;
% The safety factor for sliding failure (FS)
FD=P_A+Q_A;               % the sliding forces
FR=k_fric*(W1+W2+W3); % the resisting forces
SF_S=FR/FD;
% In the dynamic case considering eartquake forces
% The safety factor for overturning (SFe_O)
```

```
MeD=P_AD*(H/2+hb)+Q_AD*(H*2/3+hb)+MD;
MeR=MR;
SFe_O=MeR/MeD;
% The safety factor for overturning (SFe_S)
FeD=P_AD+Q_AD+FD;
FeR=FR;
SFe_S=FeR/FeD;
% 2.3. Calculation of cross-sectional effects
% Cross-sectional effects on the wall due to soil load:
for i=1:NSw
    % Axial forces for every part of the wall
    Nw(i,1)=hw*(i*H/NSw)*gama_conc;
    % Shear forces for every part of the wall
    Vw(i,1)=Ka*gama_soil*(i*H/NSw)^2*0.5+Ka*q*(i*H/NSw);
    % Bending moments for every part of the wall
    Mw(i,1)=Ka*gama_soil*(i*H/NSw)^2*0.5*(i*H/NSw^2)+Ka*q*(i*H/
NSw)*(i*H/(2*NSw));
end
% Cross-sectional effects on the base due to soil load:
L=(B2+hw+B1)*qsmin/(qsmax-qsmin); % parameter to be used in
proportion to find the base stress.
for l=1:NSb
    % Shear forces for every part of the base
    Vb(l,1)=(q+H*gama_soil+hb*gama_conc)*(l*B2/NSb)-
qsmin*(l*B2/NSb)*0.5-(qsmin*(L+l*B2/NSb)/L)*(l*B2/NSb)*0.5;
    % Bending moments for every part of the base
    Mb(l,1)=((q+H*gama_soil+hb*gama_conc)*(l*B2/NSb)^2)/2-
(qsmin*(l*B2/NSb)^2)/3-((qsmin*(L+l*B2/NSb)/L)*(l*B2/NSb)^2)/6;
end
% Additional effects due to eartquake forces on the wall:
for i=1:NSw
    Vwe(i,1)=double(int(P_ad,z,0,i*H/NSw))+K_ad*q*(i*H/NSw)+(K_
ad*q*i/NSw)*(H-i*H/NSw);
    x=double(int(P_ad*z,z,0,i*H/NSw)/int(P_ad,z,0,i*H/NSw));
    Mwe(i,1)=double(int(P_ad,z,0,i*H/NSw))*((i*H/NSw)-
x)+K_ad*q*(i*H/NSw)*(2*H*i/(NSw*3))+(K_ad*q*i/NSw)*(H-(i*H/
NSw))*(H*i/(NSw*3));
end
% Additional effects due to eartquake forces on the base:
L2=B2+hw+B1;
for l=1:NSb
    if B2*l/NSb<L2/2
         Vbe(l,1)=(1-(2*B2*l/NSb)/L2)*(-qs_add2)*(B2*l/
NSb)*0.5+(-qs_add2)*(B2*l/NSb)*0.5;
         Mbe(l,1)=(1-(2*B2*l/NSb)/L2)*(-qs_add2)*(B2*l/
NSb)^2/6+(-qs_add2)*(B2*l/NSb)^2/3;
    else
```

```
        Vbe(l,1)=(-qs_add2)*L2/2*0.5-(qs_add1*(B2*l/NSb-L2/2)/
(L2/2))*(B2*l/NSb-L2/2)*0.5;
        Mbe(l,1)=(-qs_add2)*L2/2*0.5*(l*B2/NSb-L2/2+2*L2/6)-
(qs_add1*(B2*l/NSb-L2/2)/(L2/2))*(B2*l/NSb-L2/2)^2/6;
    end
end
% 2.4. RC design of the cantilever retaining wall
% Determination of bearing design moments
% NOTE: Cross-sectional effects resulting from earthquake
effects shall be reduced by
%       Ra = 1.5 coefficient in cross-sectional calculation.
ME=[Mwe;Mbe]/1.5;
MH=[Mw;Mb];
MU1=ME+MH;    % Moment combination 1
MU2=1.6.*MH; % Moment combination 2
NG=Nw;
for r=1:NSw+NSb
    Md(r,1)=max(MU2(r,1), MH(r,1)+ME(r,1));
    % Depth of equivalent rectangular stress blocks
    a=d(r)*1000-sqrt((d(r)*1000)^2-2.*Md(r,1)*10^3/
(0.85*fcd*b));
    % Required reinforcement area [mm^2]
    As(r)=(0.85*fcd*b*1000*a/fyd);
end
smax=(1.5).*hf*1000; % upper limit vector of rebar spacing [m]
for i=1:NSw+NSb
    % Control of the minimum required reinforcement areas
    if As(i)<romin*(b*1000)*d(i)*1000
        As(i)=romin*(b*1000)*d(i)*1000;
    end
    for j=1:10
        % Generation of diameter of rebars
        R=10+j*2;
        % Generation of diameter of rebar spacings
        s=10*fix((pi*R^2/4*(1000/As(i)))/10);
        % Control of the rebar spacings considering ultimate
limits
        if s>smax(i) || s>200
            s=min(smax(i),200);
        end
        % Store these values to matrix
        D1(j,1)=R;
        D1(j,2)=s;
        D1(j,3)=(pi*R^2/4*(1000/s));
    end
    [k,t]=min(D1(:,3));
    D(i,:)=D1(t,:);
end
% Determination of shear design forces
```

```
% NOTE: Cross-sectional effects resulting from earthquake
effects shall be reduced by
%       Ra = 1.5 coefficient in cross-sectional calculation.
VE=[Vwe;Vbe]/1.5;
VH=[Vw;Vb];
VU1=VE+VH;   % Shear force combination1
VU2=1.6.*VH; % Shear force combination2
for r=1:NSw+NSb
    Vd(r,1)=max(VU2(r,1),VH(r,1)+VE(r,1)); % Shear design force
end
Vcr=0.65*(fctd*1000)*b*d;  % Shear foce capacity of concrete
% 3. OBJECTIVE FUNCTION
% Cost of concrete:
COSTC=(hw*H+hb*(B1+B2+hw))*b*Cc;
% Calculations for cost of reinforcing bars:
COSTS=0;
for i=1:NSw
    COSTS=COSTS+D(i,3)*(10^-6)*(H/NSw)*gama_steel*Cs;
end
for j=NSw+1:NSw+NSb
    COSTS=COSTS+D(j,3)*(10^-6)*(B2/NSb)*gama_steel*Cs;
end
Fx=COSTC+COSTS; % Objective function: Total cost for one meter.
% 4. PART OF PENALIZATION
for i=1:NSw+NSb
    if Vcr<Vd(i,1)
        Fx=10^12;
    end
end
if qsmin<0 || qsmax>qs || SF_O<2 || SF_S<2 || SFe_O<1.3 || SFe_
S<1.3
    Fx=10^12;
end
if Z~=4
    qs_e=1.5*qs;
    if qsmax_e>=qs_e
        Fx=10^12;
    end
else
    if  qsmax_e>=qs
        Fx=10^12;
    end
end
if imag(Fx)>0 || imag(Fx)<0
    Fx=10^12;
end
% 5. NEW SOLUTION MARIX
OPT1(I,1)=hw;
OPT1(I,2)=hb;
```

```
OPT1(I,3)=B1;
OPT1(I,4)=B2;
OPT1(I,5)=Fx;
% 6. PART OF COMPARISON
if OPT(I,5)>OPT1(I,5)
    OPT(I,:)=OPT1(I,:);
end
```

Chapter 8
Optimum Design of Carbon Fiber-Reinforced Polymer for Increasing Shear Capacity of Beams:
Optimization of CFRP for RC Beams

ABSTRACT

For reinforced concrete (RC) structures, retrofit of structures are needed to be done for several situations. These situations include the renovation of structure by adding new components (floors or extension) and elimination of safety risks (resulting from unforeseen effects - forces and durability). Most retrofit methods for RC structures need destruction of existing members and hard work on increasing of existing section dimension and reinforcements. Whereas, using carbon fiber reinforced polymer (CFRP) strips is an easy option to increase the flexural moment or shear capacity of RC members without destruction. In that case, the use of the structure is provided during the application. In this chapter, the optimum design of CFRP strips is presented for increasing the insufficient shear capacity of RC beams. The design constraints are provided according to ACI-318: Building code requirements for structural concrete and ACI-440: Guide for the Design and Construction of Externally Bonded FRP Systems for Strengthening Concrete Structure.

DOI: 10.4018/978-1-7998-2664-4.ch008

INTRODUCTION

In addition to design of new reinforced concrete structures, design of existing structures for retrofit is needed. Since the first design of structures is done according to unforeseen effects and situations, retrofit of RC structures may be needed. Retrofit of structures are done for renovation or safety situation as follows:

Renovation:

- Adding a span to structure.
- Adding a floor to structure.
- Changing span lengths for increasing free space.
- Adding a non-structural component for storage or architectural renovation.

Safety:

- To consider unforeseen effect of forces in design.
- Effects resulting from unforeseen winds and earthquakes.
- Durability reasons resulting with the strength loss of existing materials.
- Blast and accidents damaging or affected the structure with vibration.

All the mentioned reasons have an internal force increase effect. Generally, retrofit of structures are done to increase the strength capacity of structure. For that reason, it involves adding new members (dampers, shear walls, additional columns and beams, structural control systems like tuned mass dampers), increasing the dimensions of existing members by adding rebars (jacketing) and adding new materials (steel profiles, wrapping of member via a composite material). As a composite, carbon fiber reinforced polymer (CFRP) is highly used and it is a non-destructive application, and the use of structure can be provided during retrofit. Since optimization of RC members is an important issue, the optimum retrofit design is also important. In this chapter, optimum design of CFRP for increasing shear force capacity of RC beams is presented.

METHODOLOGY

In the presented methodology, the objective of the optimization is to minimize the required CFRP area per meter; A, by finding optimum values of the design variables such as width; w_f, spacing; s_f and angle; β of CFRP. The aim of the application is to gain an additional shear force capacity for an RC cross section with breadth; b_w and effective depth; d. For that reason, the design constraints are related to the shear capacity, and these constraints are defined according to ACI-318: Building code requirements for structural concrete and ACI-440: Guide for the Design and Construction of Externally Bonded FRP Systems for Strengthening Concrete Structure. The T-shaped cross section and the design variables are shown as Fig. 1a for β=90 and Fig. 1b for general cases. d_f is the depth of the RC member covered with CFRP and it is formulated as Eq. (1), and h_f denotes the height of the RC slab.

$$d_f = d - h_f \tag{1}$$

The objective function is formulated as Eq. (2). The optimum design of beam is done for unit m length of beam.

$$A = \frac{w_f\left(\dfrac{2d_f}{\sin\beta} + b\right)}{s_f} \times 1000 \tag{2}$$

Figure 1. The design variables
(Kayabekir, Sayin, Bekdaş, & Nigdeli, 2018)

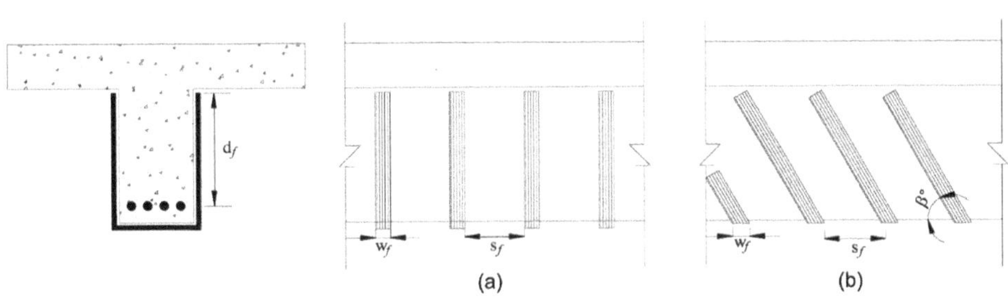

The objective function must be minimized by providing the following constraints:

$$g_1(x): s_f \leq d/4 \tag{3}$$

$$g_2(x): V_{additional} < 0.7R\frac{(2t_f w_f f_{fe})(\sin\beta + \cos\beta)d_f}{s_f + w_f} \tag{4}$$

$$g_3(x): \frac{(2t_f w_f f_{fe})(\sin\beta + \cos\beta)}{s_f + w_f} \leq \frac{2\sqrt{f_c'}b_w d}{3} - V_s \tag{5}$$

The design constants used in formulations are listed in Table 1. The optimization methodology can be given in 5 classical steps.

Step 1: Definition of design constants, ranges and algorithm parameters: In this step, all constant values are defined.

Step 2: Generation of an initial solution matrix: As the start of the optimization process, a matrix including sets of candidate solutions is defined by considering the whole solution range. In assignment of design variables, discrete variables are used for practical design. The proposed values

Table 1. The design constants

Definition	Symbol
Breadth	b_w
Height	h
Effective depth	d
Thickness of CFRP	t_f
Reduction factor	R
Thickness of slab	h_f
Comp. strength of concrete	f_c'
Effective tensile strength of CFRP	f_{fe}
Additional shear force	$V_{additional}$
Shear force capacity of rebar	V_s

for β is multiples of 5°. For dimension variables such as w_f and s_f, the proposed values are multiples of 10 mm. For all set of solutions, design objective and constraints are calculated, and the objective function is penalized with a big value for violating design constraints. For example, 10^{16} mm^2 per m can be taken as done in numerical cases.

Step 3: Generation of new solutions: New candidate solution for design variables are generated according to formulations of metaheuristic algorithms presented in Chapter 2.

Step 4: Modification of the existing solutions: All newly generated solutions are compared with the existing ones for minimization of the objective function. Better ones are stored.

Step 5: Iterative optimization: The processes given in Steps 3-4 are repeated for a maximum iteration number.

THE OPTIMUM RESULTS

The optimization results are presented for several additional shear force values as given in Kayabekir *et al.* (2018). For the values of design constants and ranges given in Table 2, the optimum results are given as Tables 3 for different additional shear force capacity. The optimization is repeated for 20 times and best (A_{best}) and mean (A_{mean}) area of CFRP, standard deviation and needed iteration values to reach to optimum results are also presented. The employed algorithm for the presented results is teaching-learning-based optimization (TLBO) (Rao, Savsani, & Vakharia, 2011).

The optimum results for both algorithms are the same. The only difference can be seen from the optimum results of design variables, but the width per spacing values are the same.

Cracks resulting from shear deformations are seen with 45° angle. In that case, positioning of members with 45° angle is the best solution as seen from the constraint given as Eq. (4), but the application of CFRP with an angle may increase the total area of CFRP strip. For that reason, the optimum angles are bigger than 45°. For comparation and validation of this theory, the optimum results are also presented for a constant β taken as 45° in Tables 4. Also, the optimum results for β=90° are presented in Table 5 since application angle is not generally considered in practical design, but a result cannot be found for 150 kN additional shear force capacity. Also, nearly 20% economy in material is provided by optimizing the application angle.

Table 2. The values of design constants and ranges

Definition	Symbol	Unit	Value
Breadth	b_w	mm	200
Height	h	mm	500
Effective depth	d	mm	450
Thickness of CFRP	t_f	mm	0.165
Reduction factor	R	-	0.5
Thickness of slab	h_f	mm	100
Comp. strength of concrete	f_c'	MPa	20
Effective tensile strength of CFRP	f_{fe}	MPa	3790
Width of CFRP	w_f	mm	10-100
Specific of CFRP	s_f	mm	0-d/4
Angle of CFRP	β	°	0-90
Additional shear force	$V_{additional}$	kN	50 (case 1) 75 (case 2) 100 (case 3) 125 (case 4) 150 (case 5)
Shear force capacity of rebar	V_s	kN	50

(Kayabekir et al., 2018)

Table 3. Optimum results

	Case 1	Case 2	Case 3	Case 4	Case 5
w_f (mm)	30	50	50	80	100
s_f (mm)	90	80	50	50	30
β (°)	65	70	65	65	70
A_{best} (mm^2)	243091.13	363432.47	486182.27	598378.18	726864.95
A_{mean} (mm^2)	243091.13	363432.47	486182.27	598378.18	726864.95
Standard deviation	$5.97x10^{-11}$	$5.97x10^{-11}$	$1.19x10^{-10}$	$1.19x10^{-10}$	$1.19x10^{-10}$
Iteration number	140	100	60	100	60

(Kayabekir et al., 2018)

Table 4. Optimum results (β=45°)

	Case 1	Case 2	Case 3	Case 4	Case 5
w_f (mm)	30	60	60	70	90
s_f (mm)	100	110	70	50	40
A_{best} (mm^2)	274603.72	419982.17	549207.45	694137.20	823811.18
Iteration number	60	60	60	60	60

(Kayabekir et al., 2018)

Table 5. Optimum results (β=90°)

	Case 1	Case 2	Case 3	Case 4	Case 5
w_f (mm)	40	40	60	90	-
s_f (mm)	80	40	30	20	-
A_{best} (mm^2)	300000	450000	600000	736363.63	-
Iteration number	60	60	60	60	-

(Kayabekir et al., 2018)

In the Appendix, a computer code written in Matlab (2018) is presented with detailed information.

REFERENCES

ACI-318. (2005). Building code requirements for structural concrete and commentary, metric version, American Concrete Institute.

American Concrete Institute Committee 440. (2006). *Guide for the Design and Construction of Structural Concrete Reinforced with FRP Bars*. American Concrete Institute.

Kayabekir, A. E., Sayin, B., Bekdaş, G., & Nigdeli, S. M. (2018, May). The factor of optimum angle of carbon fiber reinforced polymers. In *4th International Conference on Engineering and Natural Sciences (ICENS 2018)*.

Mathworks. (2018). MATLAB R2018a. The MathWorks Inc., Natick, MA.

Rao, R. V., Savsani, V. J., & Vakharia, D. P. (2011). Teaching–learning-based optimization: A novel method for constrained mechanical design optimization problems. *Computer Aided Design*, *43*(3), 303–315. doi:10.1016/j.cad.2010.12.015

APPENDIX

```
clear all
clc
format compact
%1.ENTER DATA OF PROBLEM
bw=200; % width of the beam cross-section[mm]
h=500;  % height of the beam cross-section[mm]
d=450;  % depth from the top of the section to the tension
steel reinforcement centroid[mm]
tf=0.165; % thickness of the FRP sheet on one side of the beam
R=0.5;    % ultimate tensile strength of the FRP sheet in the
direction of the principal fibers
hf=100;   % thickness of slab [mm]
fc=20;    % nominal concrete compressive strength [MPa] --->
[N/mm^2]
ffu=3790; % ultimate tensile strength of the FRP sheet in the
direction of the principal fibers [MPa] ---> [N/mm^2]
fi=0.7;   % strength reduction factor (ACI format)
% User defined algorithm parameters
maxiter=10000; % maximum iteration number
pn=20;         % population number
% Ultimate limits of design variables
% wfmin: lower limit for width of FRP strip
% wfmax: upper limit for width of FRP strip
% sfmin: lower limit for breadth of spacing of FRP strips
% sfmax: upper limit for breadth of spacing of FRP strips
% betamin: lower limit for breadth of angle between the
principal fiber orientation and the longitudinal axis of the
beam
% betamax: upper limit for breadth of angle between the
principal fiber orientation and the longitudinal axis of the
beam
wfmin=10 ;  wfmax=100;             %[mm]
sfmin=0  ;  sfmax=fix((d/4)/10)*10; %[mm]
betamin=0;  betamax=90;            %[degree]
Vneed=50000; % additinally needed shear strength
Vs=50000;    % nominal shear strength provided by steel shear
reinforcement%[N]
%2. GENERATING OF INITIAL SOLUTION MATRIX
% wf: width of FRP strip
% sf: spacing of FRP strips
% beta: angle between the principal fiber orientation and the
longitudinal axis of the beam
% Vf: nominal shear strength provided by FRP shear
reinforcement
% C : objective function
```

```
% rand: a function generating uniformly distributed
pseudorandom numbers in MATLAB
for i=1:pn % During the this loop, candidate solution vectors
including design variables are generated up to pn.
    % End of the this loop, candidate solution vectors
providing design constraint are stored in a matrix.
    %2.1. DEFINITION OF DESING VARIABLES RANDOMLY
    wf=(round((wfmin+rand*(wfmax-wfmin))/10))*10;
    sf=(round((sfmin+rand*(sfmax-sfmin))/10))*10;
    beta=(round((betamin+rand*(betamax-betamin))/5))*5;
    % 2.2. OBJECTIVE FUNCTON
    C=(2*((d-hf)/sind(beta))+bw)*wf*1000/(sf+wf);
    % 2.3. DESIGN CONSTRAINTS
    Vf=(2*tf*wf*R*ffu*(sind(beta)+cosd(beta))*(d-hf))/(sf+wf);
    % 2.4. PART OF PENALIZATION
    %NOTE: In this part,if generated any design variables
dosn't provide to regulation, objective function is penalized
with value too large.
    %With this penaltization, selection of a solution vector
including improper values of generated design variable is
prevented.
    if fi*Vf<Vneed || 2*sqrt(fc)*bw*d/3-Vs<Vf % Control of
design constraint according to ACI440
        C=10^16;                               % Penalization
with high value
    end
    % 2.5. INITIAL SOLUTION MATRIX
    % NORE: End of the each loop, a candidate solutution vector
including randomly generated design variables are stored in the
matrix named OPT
    % Thus each vector in matrix is candidate solution vector.
    OPT(1,i)=wf;
    OPT(2,i)=sf;
    OPT(3,i)=beta;
    OPT(4,i)=C;
end
% 3. GENERATING OF NEW SOLUTION MATRIX VIA TLBO
% NOTE: This part is very similar to PART 2. Differently, new
solution variables are generated according to TLBO algorithm
% and ultimate limits of variables are controlled.
% All steps during the PART3 are repeated until all iterations
are completed.
for iter=1:maxiter
    % 3.1. TEACHER PHASE
    for i=1:pn
        % 3.1.1. DEFINATION OF DESING VARIABLES
        % Values of candidate solution vector providing best/
minimum value of objective function in existing solution set or
matrix are assigned to vector named best:
```

```
        [p,r]=min(OPT(4,:));
        best(1,1)=OPT(1,r);
        best(2,1)=OPT(2,r);
        best(3,1)=OPT(3,r);
        % New values of design variables are generated
according to formulation of TLBO algorithm
        wf=wf+(round((rand*(best(1,1)-round(1+rand)*mean(O
PT(1,:))))/10))*10;
        sf=sf+(round((rand*(best(2,1)-round(1+rand)*mean(O
PT(2,:))))/10))*10;
        beta=beta+(round((rand*(best(3,1)-round(1+rand)*mean(O
PT(3,:))))/5))*5;
        % 3.1.2. CONTROL OF ULTIMATE LIMITS
        if wf>wfmax
            wf=wfmax;
        end
        if wf<wfmin
            wf=wfmin;
        end
        if sf>sfmax
            sf=sfmax;
        end
        if sf<sfmin
            sf=sfmin;
        end
        if beta>betamax
            beta=betamax;
        end
        if beta<betamin
            beta=betamin;
        end
        % 3.1.3. OBJECTIVE FUNCTION
        C=(2*((d-hf)/sind(beta))+bw)*wf*1000/(sf+wf);
        % 3.1.4. DESIGN CONSTRAINTS
        Vf=(2*tf*wf*R*ffu*(sind(beta)+cosd(beta))*(d-hf))/
(sf+wf);
        % 3.1.5. PART OF PENALIZATION
        %NOTE: In this part,if generated any design variables
don't provide ACI440 rules, objective function is penalized
with value too large.
        %With this penaltization, selection of a solution
vector including improper values of generated design variable
is prevented.
        if fi*Vf<Vneed || 2*sqrt(fc)*bw*d/3-Vs<Vf % Control of
design constraint according to ACI440
            C=10^16;                                    %
Penalization with high value
        end
        % 3.1.6. NEW SOLUTION MATRIX
```

```
        OPT1(1,i)=wf;
        OPT1(2,i)=sf;
        OPT1(3,i)=beta;
        OPT1(4,i)=C;
        % 3.1.7. PART OF COMPARISON
        % Each candidate solution vector in the new solution
matrix named OPT1 are compared with existing solution vectors
in maxtrix OPT
        % If new solution vectors provide better solution or
minimum cost, it is replaced with existing solution vectors:
        if OPT(4,i)>OPT1(4,i)  % Comparing of values of
objective functions
            OPT(:,i)=OPT1(:,i); % Updating of matrix OPT
        end
    end
    % 3.2. LEARNER PHASE
    for  i=1:pn
        %Random numbers are generated between 1 and pn to be
used in generation of new design variables:
        %Note: ceil is a command used to round a number to the
nearest integer greater than or equal to.
        v1=ceil(rand*pn);
        v2=ceil(rand*pn);
        while v1==v2
            v1=ceil(rand*pn);
            v2=ceil(rand*pn);
        end
        % New values of design variables are generated
according to randomly selected two design variables
        wf=OPT(1,i)+(round(rand*(OPT(1,v1)-
(OPT(1,v2)))/10))*10;
        sf=OPT(2,i)+(round(rand*(OPT(2,v1)-
(OPT(2,v2)))/10))*10;
        beta=OPT(3,i)+(round(rand*(OPT(3,v1)-
(OPT(3,v2)))/5))*5;
        % 3.2.2. CONTROL OF ULTIMATE LIMITS
        if beta>betamax
            beta=betamax;
        end
        if beta<betamin
            beta=betamin;
        end
        if wf>wfmax
            wf=wfmax;
        end
        if wf<wfmin
            wf=wfmin;
        end
        if sf>sfmax
```

```
            sf=sfmax;
        end
        if sf<sfmin
            sf=sfmin;
        end
        % 3.2.3. OBJECTIVE FUNCTION
        C=(2*((d-hf)/sind(beta))+bw)*wf*1000/(sf+wf);
        % 3.2.4. DESIGN CONSTRAINTS
        Vf=(2*tf*wf*R*ffu*(sind(beta)+cosd(beta))*(d-hf))/
(sf+wf);
        % 3.2.5. PART OF PENALIZATION
        if fi*Vf<Vneed || 2*sqrt(fc)*bw*d/3-Vs<Vf % Control of
design constraint according to ACI440
            C=10^16;                                %
Penalization with high value
        end
        % 3.2.6. NEW SOLUTION MATRIX
        OPT1(1,i)=wf;
        OPT1(2,i)=sf;
        OPT1(3,i)=beta;
        OPT1(4,i)=C;
        % 3.2.7. PART OF COMPARISON
        if OPT(4,i)>OPT1(4,i)  % Comparing of values of
objective functions
            OPT(:,i)=OPT1(:,i); % Updating of existing matrix
OPT
        end
    end
    [mi,ni]=min(OPT(4,:));
    mi
end
```

Chapter 9
Optimum Design of Post-Tensioned Axially-Symmetric Cylindrical Reinforced Concrete Walls

ABSTRACT

In this chapter, an optimization methodology for design of post-tensioned axially symmetric cylindrical reinforced concrete (RC) walls is presented. The objective of optimization is the minimization of total material cost of the wall including concrete, reinforced bars, post-tensioned cables, and form work required for wall and application of the post-tensioning. The optimized values are wall thickness, compressive strength of the concrete, locations and intensities of the post-tensioned loads, the diameter of the reinforcement bars (rebars), and distance between rebars. The optimization process employs the superposition method (SPM) for the analyses of the wall, and the design constraints are defined according to ACI-318: Building code requirements for structural concrete.

INTRODUCTION

For the storage of liquids, water tanks are formed by using analyzing and designing axially symmetric cylindrical walls. Since the storage is a liquid that can lead easily, serviceability requirements including the limitation of

DOI: 10.4018/978-1-7998-2664-4.ch009

cracking under applied loads must be critically considered in addition to strength factors. All serviceability and structural state limits (stress factors) are formulated and ruled in design regulations such as ACI 318: Building code requirements for structural concrete. To provide all serviceability and stress conditions, unphysical solutions with very large wall thickness, big diameter and dense spacing of rebars may be needed. In this situation, water tank may not be economic. To decrease costs and provide better serviceability, the application of post-tension systems via cables can be considered. An optimized post-tensioning forces combination can reduce the magnitude of the internal forces, and due to that, slender container wall designs, decrease in the amount of rebars and reduction in the number and size of cracks can be provided.

The optimum design of RC liquid tanks has been investigated by only considering the optimum design of RC design without post-tensioning (Adidam & Subramanyam 1982; Saxena, Sharma, & Mohan, 1987; Thevendran & Thambiratnam 1988; Tan, Thevendran, & Gupta 1993; Barakat & Altoubat 2009; El Ansary, El Damatty, & Nassef, 2011; Bekdaş, 2014). The optimum design of reinforced concrete (RC) cylindrical walls with post-tensioning forces was also optimized by using harmony search (HS) (Bekdaş, 2015) and other algorithms such as Harmony Search (HS), Flower Pollination Algorithm (FPA), Teaching Learning Based Optimization (TLBO), Jaya Algorithm (JA) including hybrid ones (Bekdaş, 2018, 2019). In this chapter, the methodology developed by Bekdaş (2015) is presented.

OPTIMIZATION DESIGN OF POST-TENSIONED AXIALLY SYMMETRIC CYLINDRICAL RC WALLS

Formulation of Cylindrical Walls

The equations of cylindrical walls can be obtained from the element presented as Fig. 1. According to the symmetry of the element, the shearing forces on the membrane vanish and forces; N_φ are constant along the circumference. Also, only transverse shearing force is Q_x. The twisting moments also vanish, and the bending moments; M_φ are constant along the circumference.

Under this condition, it is possible to derive three of the six equations of the equilibrium of the element. If only pressure normal to the surface is

Figure 1. An element of cylindrical wall for derivation of the equations (Bekdaş, 2015)

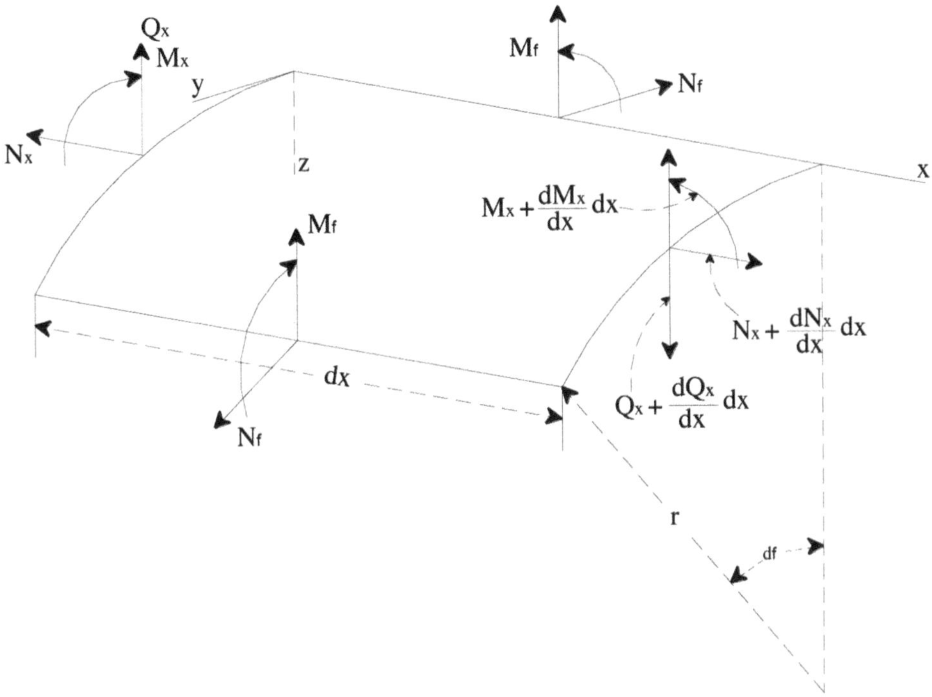

assumed for the external forces are assumed, equations for the equilibrium of the moment and forces can be written as Eq. (1).

$$\frac{dN_x}{dx} r dx d\phi = 0, \ \frac{dQ_x}{dx} r dx d\phi + N_\phi dx d\phi + q r dx d\phi = 0 \text{ and}$$

$$\frac{dM_x}{dx} r dx d\phi - Q_x r dx d\phi = 0 \tag{1}$$

Since the N_x forces are constant, Eq. (1) can be simplified as follows:

$$\frac{dQ_x}{dx} + \frac{1}{r} N_\phi = -q \text{ and } \frac{dM_x}{dx} - Q_x = 0 \tag{2}$$

Eq. (2) has three unknown quantities such as N_φ, Q_x and M_x. If the u and w denote the displacements in the x and z directions, the unit strains in the corresponding directions are as follows:

$$\varepsilon_x = \frac{du}{dx} \text{ and } \varepsilon_\phi = -\frac{w}{r} \tag{3}$$

By inserting Eq. (3) into equations of N_x and N_φ, the following equations are obtained:

$$\begin{aligned} N_x &= \frac{Eh}{1-\nu^2}(\varepsilon_x + \nu\varepsilon_\phi) = \frac{Eh}{1-\nu^2}\left(\frac{du}{dx} - \nu\frac{w}{r}\right) = 0\,and \\ N_\phi &= \frac{Eh}{1-\nu^2}(\varepsilon_\phi + \nu\varepsilon_x) = \frac{Eh}{1-\nu^2}\left(-\frac{w}{r} + \nu\frac{du}{dx}\right) \end{aligned} \tag{4}$$

υ denotes Poisson ratio and h represents the thickness of the wall. From the expressions given as Eq. (4), the following equations are obtained:

$$\frac{du}{dx} = \nu\frac{w}{r} \tag{5}$$

$$N_\phi = -\frac{Ehw}{r} \tag{6}$$

Due to symmetry, no change in curvature in the circumferential direction occurs. For that reason, the curvature in the x direction is equal to $-dw^2/dx^2$. By using the same moment expressions for plates, Eq. (7) is obtained.

$$M_\phi = \nu M_x \text{ and } M_x = -D\frac{dw^2}{dx^2} \tag{7}$$

D is the flexural rigidity of the shell or plate, and it is formulated as Eq. (8). By inserting Eqs. (5)-(7) into Eq.(2), the differential equation given as Eq. (9) is found.

$$D = \frac{Eh^3}{12(1-\nu^2)} \tag{8}$$

$$\frac{d^2}{dx^2}\left(D\frac{d^2w}{dx^2}\right)+\frac{Eh}{r^2}w=q \tag{9}$$

If the thickness of the shell is constant, Eq. (9) takes the following simplified form:

$$D\frac{d^4w}{dx^4}+\frac{Eh}{r^2}w=q \tag{10}$$

By using the notation given as Eq. (11), the final equation takes the form given as Eq. (11).

$$\beta^4=\frac{Eh}{4r^2D}=\frac{3(1-\nu^2)}{r^2h^2} \tag{11}$$

$$\frac{d^4w}{dx^4}+4\beta^4w=\frac{q}{D} \tag{12}$$

Eq. (12) is similar to the differential equation of a beam supported by a continuous elastic foundation, subjected a uniform load of intensity q, and having a flexural rigidity defined as D (Timoshenko & Young, 1962). The general solution of Eq. (12) is as follows:

$$w=e^{\beta x}(C_1\cos\beta x+C_2\sin\beta x)+e^{-\beta x}(C_3\cos\beta x+C_4\sin\beta x)+f(x) \tag{13}$$

f(x) is a particular solution of the equation, and C_1, C_2, C_3 and C_4 represent integration constants based on the boundary conditions.

Until now, various alternative methods including superposition method was proposed for analyses of cylindrical walls or beams. In the presented methodology of optimization, the analyses were conducted by using the superposition method (SPM). SPM has been developed by Hetenyi (1946) for the design solution of beams and it can be also accepted for the solution of walls because of the similarity of differential equations. The analysis process of the superposition methods is explained in three steps.

Step 1: Calculation of All Internal Forces by Assuming Wall as Infinite Long

As the advantage of the method comparing to other methods (Billington, 1965; Selvadurai, 1979; Ghali, 1979; Timoshenko & Woinowsky-Krieger, 1984; Kelkar & Sewell, 1987; Melerski, 2000; Ventsel & Krauthammer, 2001), determining the integration constant for beams (or walls) of infinite length van be shown. First, the infinite long wall analysis principles must be mentioned before the explanation of the method. If an infinite long wall is considered in both directions subjected to a single concentrated load P at the midpoint as shown in Fig. 2, it is possible to assume that the deflection of the wall approaches zero in an infinite distance from the application point of the load. As symbolic, if x→∞, then w→0. According to this assumption, terms with $e^{\beta x}$ is vanished from Eq. (13), and the form presented as Eq. (14) obtained.

$$w = e^{-\beta x}(C_3 \cos \beta x + C_4 \sin \beta x) \tag{14}$$

According to symmetry, the expressions $[dw/dx]_{x=0}=0$ and $[Q]_{x=0}= -P/2$ are used to modify Eq. (14) to obtain Eq. (15).

$$w = \frac{P}{8\beta^3 D} e^{-\beta x}(\cos \beta x + \sin \beta x) \tag{15}$$

As presented in Eq. (16), the expressions for angular displacement (θ), moment (M) and shear force (Q) are written by taking the derivatives of w with respect to x, and use the notations formulated as Eq. (17).

$$w = \frac{P}{8\beta^3 D} A_{\beta x},\ \theta = -\frac{P}{4\beta^2 D} B_{\beta x},\ M = \frac{P}{4\beta} C_{\beta x},\ \text{and } Q = -\frac{P}{2} D_{\beta x}, \tag{16}$$

Figure 2. Infinite-length wall under the influence of a concentrated load (Bekdaş, 2015)

$$
\begin{aligned}
A_{\beta x} &= e^{-\beta x}(\cos \beta x + \sin \beta x) \\
B_{\beta x} &= e^{-\beta x} \sin \beta x \\
C_{\beta x} &= e^{-\beta x}(\cos \beta x - \sin \beta x) \\
D_{\beta x} &= e^{-\beta x} \cos \beta x
\end{aligned} \tag{17}
$$

Similarly, the expressions of an infinitely long wall subjected to a single concentrated moment $M_{0,}$ at the midpoint is written as Eq. (18).

$$
w = \frac{M_0}{4\beta^2 D} B_{\beta x},\ \theta = \frac{M_0}{4\beta D} C_{\beta x},\ M = \frac{M_0}{2} D_{\beta x},\ \text{and}\ Q = -\frac{P}{2} D_{\beta x}, \tag{18}
$$

In application of SPM procedure, a wall subjected to a given loading condition is considered as shown in Fig. 3a. In the first step, the internal forces of the wall are calculated by the assumption of the wall as an infinitely long one as given in Fig. 3b, where the internal forces approach to zero at distances that are infinitely far from the loads. Thus, some internal forces occur at the ends of the wall, and a moment M_A and a shear force Q_A at point A, and similarly, at point B (a moment M_B and a shear force Q_B) is seen. These forces and moments must be zero according to boundary condition of the wall shown as Fig. 3a.

Step 2: Determination of the End-conditioning Forces

The boundary condition is provided by using P_{0A}, M_{0A}, P_{0B} and M_{0B} forces, and these forces are called as end-conditioning forces as shown in Fig 3c. By using the end-conditioning forces, -M_A and -Q_A at point A, and -M_B and -Q_B at point B must be obtained. In that case, the behavior of the infinitely long wall given as Fig. 3c will be the same as the wall given as Fig. 3a.

By considering the four conditions of the two ends, the values of the end-conditioning forces are written as Eq. (19) by using the formulations given as Eqs. (16) and (17). $A_{\beta l}$, $B_{\beta l}$, $C_{\beta l}$ and $D_{\beta l}$ are the values of the functions given as Eq. (17) when $x=l$.

$$
M_A + \frac{P_{0A}}{4\beta} + \frac{P_{0B}}{4\beta} C_{\beta l} + \frac{M_{0A}}{2} + \frac{M_{0B}}{2} D_{\beta l} = 0,
$$

Figure 3. SPM for walls with both ends free
(Bekdaş, 2015)

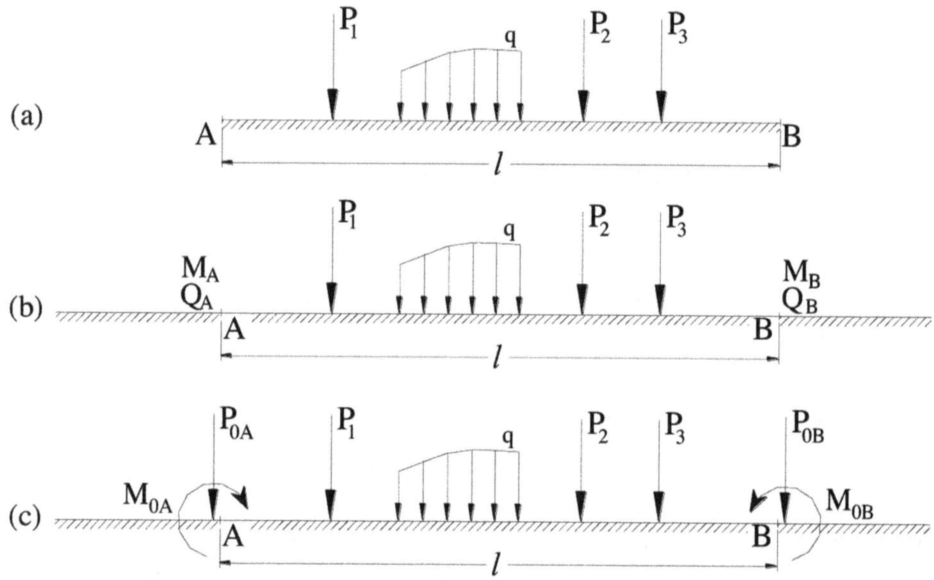

$$Q_A - \frac{P_{0A}}{2} + \frac{P_{0B}}{2} D_{\beta l} - \frac{\beta M_{0A}}{2} + \frac{\beta M_{0B}}{2} A_{\beta l} = 0,$$

$$M_B + \frac{P_{0A}}{4\beta} C_{\beta l} + \frac{P_{0B}}{4\beta} + \frac{M_{0A}}{2} D_{\beta l} + \frac{M_{0B}}{2} = 0,$$

and

$$Q_B - \frac{P_{0A}}{2} D_{\beta l} + \frac{P_{0B}}{2} - \frac{\beta M_{0A}}{2} A_{\beta l} - \frac{\beta M_{0B}}{2} = 0 \tag{19}$$

Step 3: The Superposition Process

In the final step, the internal forces are obtained by superposing the values found according to the end-conditioning forces and the loading conditions. In this chapter, the equations are presented for walls with both ends free, and the expressions are differently derived for different boundary conditions.

Figure 4. Model of cylindrical wall used in optimization process (Bekdaş, 2015)

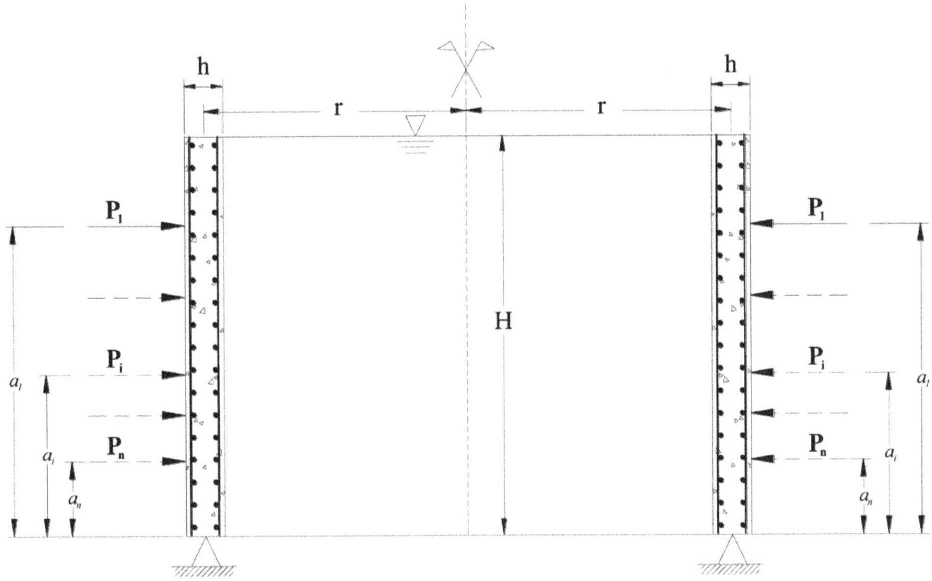

The Optimization Methodology

The optimized cylindrical wall model is given as Fig. 4. P_1,…., P_i,…. and P_n represent n post-tensioning loads with application location such as a_1,…., a_i,… and a_n. h, H and r are thickness, height and radius of the wall, respectively.

The main aim of optimization processes is to determine design variables presented in Table 1 by minimizing the objective function providing the design constraints presented as Table 2 defined according to ACI-318: Building code requirements for structural concrete. The number of design variables is $2n+6$ because the number of post-tensioning loads (n) is also randomly chosen for all iterations of optimization. In Table 2, the flexural moment, the shear force, the axial tension, the area of hoop reinforcing bars, the yield strength of steel, the area of vertical reinforcing bars and the area of horizontal reinforcing bars are denoted with M_u, V_u, T_u, A_s^{hoop}, f_y, A_{sv} and A_{sh}, respectively.

The total cost of materials including concrete, rebars, post-tensioned cables and form work is the objective function (min(f(X))) of the problem as given in Eq. (20). In Eq. (20), C_s, C_{pt} and C_{fw} are the unit cost of concrete, steel, post-tensioning and formwork, respectively. These values are design constants. The value of unit cost of concrete ($C_c(x_{2n+2})$) is defined according

Table 1. Design variables

Description	Design Variable
Post-tensioning loads	$x_1=P_1, x_2=P_2, \ldots x_n=P_n$
Locations of post-tensioning loads	$x_{n+1}=a_1, x_{n+2}=a_2, \ldots x_{2n}=a_n$
Thickness of the wall, h	$x_{2n+1}=h$
Compressive strength of the concrete, f'_c	$x_{2n+2}=f'_c$
Diameter of vertical reinforcing bars, ϕ_v	$x_{2n+3}=\phi_v$
Diameter of horizontal reinforcing bars, ϕ_h	$x_{2n+4}=\phi_h$
Distance between vertical reinforcing bars, S_v	$x_{2n+5}=S_v$
Distance between horizontal reinforcing bars, S_h	$x_{2n+6}=S_h$

Table 2. Constraints on strength and dimensions of wall

Description	Constraints
Flexural strength capacity, M_d	$g_1(X)$: $M_d \geq M_u$
Shear strength capacity, V_d	$g_2(X)$: $V_d \geq V_u$
The axial tension, T	$g_3(X)$: $T=\phi A_s^{hoop} f_y \geq T_u$
Minimum steel area, A_{smin}	$g_4(X)$: $A_{sv} \geq A_{smin}$ and $A_{sh} \geq A_{smin}$
Maximum crack width, w_{max}	$g_5(X)$: $w_{max} \leq 0.1$ mm
Maximum steel bars spacing, S_{max}	$g_6(X)$: $S_v \leq S_{max}$ and $S_h \leq S_{max}$
Minimum steel bars spacing, S_{min}	$g_7(X)$: $S_v \geq S_{min}$ and $S_h \geq S_{min}$
Minimum concrete cover, c_{cmin}	$g_8(X)$: $c_{cmin} \geq 40$ mm

to local market prices respect to a design variable; compressive strength of concrete. The cost of concrete is per unit volume, the steel cost is per unit weight, and the cost of the formwork is per unit area. In Eq. (21), $V_c\,(x_{2n+1})$ represents the wall volume.

$$\min f(X) = C_c(x_{2n+2})V_c(x_{2n+1}) + C_s W_s(x_{2n+3}, x_{2n+4}, x_{2n+5}, x_{2n+6}) + C_{pt} W_{pt}(x_1, \ldots x_{2n}) + C_{fw} A_{fw} \tag{20}$$

$$V_c(x_{2n+1}) = \pi H\left(\left(r + \frac{h}{2}\right)^2 - \left(r - \frac{h}{2}\right)^2\right) \tag{21}$$

The total weight of the reinforcements includes inner, outer hoop and vertical reinforcements; $W_s(x_{2n+3}, x_{2n+4}, x_{2n+5}, x_{2n+6})$ are formulated as Eq. (22). Eq. (23) represents the total weight of the post-tensioning cables; $W_{pt}(x_1,\ldots x_{2n})$.

$$W_s(x_{2n+3}, x_{2n+4}, x_{2n+5}, x_{2n+6}) = \frac{\pi\varphi_v^2}{4}\left(\frac{4\pi r}{S_v}\right)H\gamma_s + \frac{\pi\varphi_h^2}{4}\left(\frac{2H}{S_h}\right)2\pi r\gamma_s \quad (22)$$

$$W_{pt}(x_1,\ldots x_{2n}) = \frac{\pi\sum_{i=1}^{n}(\varphi_{pt})_i^2}{4}2\pi r\gamma_s \quad (23)$$

γ_s and ϕ_{pt} represents the density of the steel and the diameter of the post-tensioning cables, respectively. A_{fw} is the sum of wall external and internal surface area as formulated in Eq. (24).

$$A_{fw} = 2\pi H(2r) \quad (24)$$

THE OPTIMUM RESULTS FOR NUMERICAL CASES

The optimum results of five different wall models with different heights was done by Bekdaş (2015) employing harmony search algorithm (Geem, Kim, & Loganathan, 2001). The results of that study are presented in this chapter. The penalty factor for a solution violating a design constraint is the multiplication of calculated objection function with 10^{10} for elimination. To realize that the post-tensioning forces are required in an economic design, post-tensioning forces are used with a 50% probability in the generation of a new set of candidate solutions. All presented results are for maximum $2x10^6$ iterations.

The radius of the wall and the Poisson's ratio are taken as 18.25 m and 0.15, respectively. The elasticity modulus is found according to Eq. (25). The range of compressive strength of concrete is between 20 MPa - 50 MPa. The range of the height of the wall is from 0.05 m-2.0 m. Post-tensioning loads were randomized between 0 kN/m and $2\gamma H$ kN/m (per meter) values. The density of the water (γ) is 9.81 kN/m^3. The ranges of rebars sizes are between 8 mm-50 mm. Material cost was taken as \$40/m^3 for concrete with 20 MPa compressive strength. The cost concrete is increased by \$5 for every additional 5 MPa strength. Steel rebar price is \$315/ton, and formwork is \$5/m^2. These costs are taken according to the Ministry of Environment and Urbanization of Turkey. The tensile strength of the post-tensioning cables is f_{pu}=1725 MPa, and maximum allowable stress is $0.7f_{pu}$. For the long-term effects, 26% strength loss in post-tensioning cables must be taken into consideration. The post-tensioning cable price is \$2900/ton including shipping and all other application costs.

$$E = 4700\sqrt{f_c'} \text{ (MPa)}. \tag{24}$$

The optimum results for 9 cases of are presented in Table 3. The last 4 cases include only RC design optimization and contain no post tensioning cables. The optimum intensity and location of post-tensioning forces are also given in different tables such as Table 4 and Table 5, respectively

As seen from the optimum results, post-tensioning is not an optimum option for the example 1. For the second example, a little effect on cost can be seen by post-tensioning force when the optimum results of example 2 and 6 are compared. By the increase of height of the tank, the significant advantage of post-tensioning can be clearly seen. Especially for the example 9, the optimum thickness of the wall is more than 3 times of the case with post-tensioning. By using post-tensioning, the optimum thickness remains

Table 3. Optimum results

Example No	H (m)	h (m)	Compressive Strength of Concrete	Cost (\$)
1	4.0	0.20	20	*11115.9*
2	6.0	0.25	20	*22662.2*
3	8.0	0.20	30	*36748.5*
4	10.0	0.25	20	*53242.6*
5	12.0	0.25	30	*72144.8*
6	6.0	0.35	20	*22683.0*
7	8.0	0.50	20	*37276.2*
8	10.0	0.65	20	*55488.6*
9	12.0	0.80	20	*77204.2*

(Bekdaş, 2015)

Table 4. Post-tensioned loads intensities

Example No	Load Intensity, P_i (kN)									
	1	2	3	4	5	6	7	8	9	10
1	-	-	-	-	-	-	-	-	-	-
2	0.93	9.45	14.48	9.68	0.95	0.82	0.09	6.56	9.52	8.04
3	50.18	6.52	6.42	10.97	10.10	68.40	8.77	38.72	5.52	13.92
4	6.71	3.99	56.72	85.36	80.80	10.27	5.75	67.48	18.42	34.13
5	19.21	79.02	0.02	158.85	22.29	1.41	100.23	82.08	16.29	54.78

(Bekdaş, 2015)

Table 5. Locations of post-tensioning loads

Example No	Location of Loads, a_i (m)									
	1	2	3	4	5	6	7	8	9	10
1	-	-	-	-	-	-	-	-	-	-
2	4.25	5.73	1.23	3.89	4.89	4.62	4.43	3.17	1.56	3.87
3	2.11	7.79	5.56	6.22	4.43	3.39	4.40	1.38	6.89	4.43
4	7.71	7.80	4.43	1.06	2.51	7.06	5.33	3.95	5.83	1.86
5	9.33	1.18	8.22	3.53	6.50	8.60	5.21	2.08	8.86	5.86

(Bekdaş, 2015)

in the range between 0.2 m and 0.25 m. As summary, post-tensioning have 0.09%, 1.42%, 4.05% and 6.55% cost advantage for tanks with 6 m, 8 m, 10 m and 12 m height, respectively.

REFERENCES

ACI-318. (2005). Building code requirements for structural concrete and commentary, metric version. American Concrete Institute, 2005.

Adidam, S. R., & Subramanyam, A. V. (1982). Optimum design of reinforced concrete water tanks. *Journal of the Structural Division, 108*(6), 1219–1231.

Barakat, S. A., & Altoubat, S. (2009). Application of evolutionary global optimization techniques in the design of RC water tanks. *Engineering Structures, 31*(2), 332–344. doi:10.1016/j.engstruct.2008.09.006

Bekdas, G. (2014). Optimum design of axially symmetric cylindrical reinforced concrete walls. *Structural Engineering and Mechanics*, *51*(3), 361–375. doi:10.12989em.2014.51.3.361

Bekdaş, G. (2015). Harmony search algorithm approach for optimum design of post-tensioned axially symmetric cylindrical reinforced concrete walls. *Journal of Optimization Theory and Applications*, *164*(1), 342–358. doi:10.100710957-014-0562-2

Bekdaş, G. (2018). New improved metaheuristic approaches for optimum design of posttensioned axially symmetric cylindrical reinforced concrete walls. *Structural Design of Tall and Special Buildings*, *27*(7). doi:10.1002/tal.1461

Bekdaş, G. (2019). Optimum design of post-tensioned axially symmetric cylindrical walls using novel hybrid metaheuristic methods. *Structural Design of Tall and Special Buildings*, *28*(1). doi:10.1002/tal.1550

Billington, D. P. (1965). *Thin shell structures*. New York: McGraw-Hill.

El Ansary, A. M., El Damatty, A. A., & Nassef, A. O. (2011). A coupled finite element genetic algorithm for optimum design of stiffened liquid-filled steel conical tanks. *Thin-walled Structures*, *49*(4), 482–493. doi:10.1016/j.tws.2010.12.006

Geem, Z. W., Kim, J. H., & Loganathan, G. V. (2001). A new heuristic optimization algorithm: harmony search. *simulation, 76*(2), 60-68.

Ghali, A. (1979). *Circular Storage Tanks and Soils*. London, UK: Spon Ltd.

Hetenyi, M. (1946). *Beams on Elastic Foundation*. Ann Arbor, MI: The University of Michigan Press.

Kelkar, V. S., & Sewell, R. T. (1987). *Fundamentals of the Analysis and Design of Shell Structures*. Englewood Cliffs, NJ: Prentice-Hall.

Melerski, E. D. (2000). *Design Analysis of Beams, Circular Plates and Cylindrical Tanks on Elastic Foundations*. London, UK: Taylor & Francis Group.

Saxena, M., Sharma, S. P., & Mohan, C. (1987). Cost optimization of Intze tanks on shafts using nonlinear programming. *Engineering Optimization*, *10*(4), 279–288. doi:10.1080/03052158708902543

Selvadurai, A. P. S. (1979). *Elastic Analysis of Soil-Foundation Interaction*. Amsterdam, The Netherlands: Elsevier Scientific Publishing.

Tan, G. H., Thevendran, V., Gupta, N. D., & Thambiratnam, D. P. (1993). Design of reinforced concrete cylindrical water tanks for minimum material cost. *Computers & Structures*, *48*(5), 803–810. doi:10.1016/0045-7949(93)90501-4

Thevendran, V., & Thambiratnam, D. P. (1988). Minimum weight design of conical concrete water tanks. *Computers & Structures*, *29*(4), 699–704. doi:10.1016/0045-7949(88)90382-3

Timoshenko, S., & Young, D. H. (1962). Strength of Materials 4th Edition-Part II. D. Van Nostrand Company, New Jersey.

Timoshenko, S. P., & Woinowsky-Krieger, S. (1984). *Theory of Plates and Shells* (2nd ed.). New York: McGraw-Hill.

Ventsel, E., & Krauthammer, T. (2001). *Thin Plates and Shells: Theory, Analysis and Applications*. New York: Marcel Dekker. doi:10.1201/9780203908723

Related Readings

To continue IGI Global's long-standing tradition of advancing innovation through emerging research, please find below a compiled list of recommended IGI Global book chapters and journal articles in the areas of structural design, architecture, and concrete structures. These related readings will provide additional information and guidance to further enrich your knowledge and assist you with your own research.

Afonso, J. M. (2018). The Power of Monsanto's Stone: Contribution to the Study of the Sustainable Adaptative Strategies. In I. Rosa, J. Lopes, R. Ribeiro, & A. Mendes (Eds.), *Handbook of Research on Methods and Tools for Assessing Cultural Landscape Adaptation* (pp. 122–152). Hershey, PA: IGI Global. doi:10.4018/978-1-5225-4186-8.ch006

Amador, H. C. (2018). Contemporary Urbanizations in Public Water Reservoirs: Floating Villages of Alqueva Reservoir. In I. Rosa, J. Lopes, R. Ribeiro, & A. Mendes (Eds.), *Handbook of Research on Methods and Tools for Assessing Cultural Landscape Adaptation* (pp. 249–275). Hershey, PA: IGI Global. doi:10.4018/978-1-5225-4186-8.ch010

Amoruso, G. (2015). The Image of Historic Urban Landscapes: Representation Codes. In S. Brusaporci (Ed.), *Handbook of Research on Emerging Digital Tools for Architectural Surveying, Modeling, and Representation* (pp. 550–578). Hershey, PA: IGI Global. doi:10.4018/978-1-4666-8379-2.ch019

Amoruso, G. (2017). Figuring Out the Interiors through the Geometric Tools of Representation: The Illusory Cast of Design. In L. Crespi (Ed.), *Design Innovations for Contemporary Interiors and Civic Art* (pp. 286–306). Hershey, PA: IGI Global. doi:10.4018/978-1-5225-0666-9.ch016

Amoruso, G. (2018). Geometry, Shape, and Typological Adaptation of Lighthouses Within the Italian Environment. In D. D'Uva (Ed.), *Handbook of Research on Form and Morphogenesis in Modern Architectural Contexts* (pp. 310–328). Hershey, PA: IGI Global. doi:10.4018/978-1-5225-3993-3.ch015

Anzani, A., & Caramel, C. (2017). Design and Restoration: A Holistic Approach. In L. Crespi (Ed.), *Design Innovations for Contemporary Interiors and Civic Art* (pp. 241–257). Hershey, PA: IGI Global. doi:10.4018/978-1-5225-0666-9.ch013

Ayiter, E. (2018). Playing With Text in Space. In M. Khosrow-Pour, D.B.A. (Ed.), Enhancing Art, Culture, and Design With Technological Integration (pp. 114-130). Hershey, PA: IGI Global. doi:10.4018/978-1-5225-5023-5.ch006

Balsas, C. J. (2018). A Novel Approach to Studying Cultural Landscapes at the Watershed Level. In I. Rosa, J. Lopes, R. Ribeiro, & A. Mendes (Eds.), *Handbook of Research on Methods and Tools for Assessing Cultural Landscape Adaptation* (pp. 221–248). Hershey, PA: IGI Global. doi:10.4018/978-1-5225-4186-8.ch009

Baptista, J. L., Ribeiro, J. T., & Henriques, C. D. (2018). Construction Information Map: Support for Sustainable Architecture Projects in Developing Countries – Angola Case Study. In I. Rosa, J. Lopes, R. Ribeiro, & A. Mendes (Eds.), *Handbook of Research on Methods and Tools for Assessing Cultural Landscape Adaptation* (pp. 323–345). Hershey, PA: IGI Global. doi:10.4018/978-1-5225-4186-8.ch013

Barbato, D. (2015). A Possible Method for 3D Photo Scanning Speed Survey: The Case Study of the Bridge of Saint Angelo above Auso. In S. Brusaporci (Ed.), *Handbook of Research on Emerging Digital Tools for Architectural Surveying, Modeling, and Representation* (pp. 361–381). Hershey, PA: IGI Global. doi:10.4018/978-1-4666-8379-2.ch012

Barneche-Naya, V., & Hernández-Ibáñez, L. A. (2018). A Vitruvian-Inspired Theoretical Framework for Architecture in Virtual Worlds. In M. Khosrow-Pour, D.B.A. (Ed.), Enhancing Art, Culture, and Design With Technological Integration (pp. 152-168). Hershey, PA: IGI Global. doi:10.4018/978-1-5225-5023-5.ch008

Benli, G. (2015). The Advantages of Using Laser Scanners in Surveying in Protected Sites: A Case Study in Historical Peninsula in Istanbul. In S. Brusaporci (Ed.), *Handbook of Research on Emerging Digital Tools for Architectural Surveying, Modeling, and Representation* (pp. 382–402). Hershey, PA: IGI Global. doi:10.4018/978-1-4666-8379-2.ch013

Benli, G., & Güner, A. F. (2017). Two Unique Protected Sites with a Modern Heritage in Historical Peninsula in Istanbul. In G. Koç, M. Claes, & B. Christiansen (Eds.), *Cultural Influences on Architecture* (pp. 130–156). Hershey, PA: IGI Global. doi:10.4018/978-1-5225-1744-3.ch005

Bertron, A. (2017). On Exhibition Graphics Understood as a Space-Related Interface. In L. Crespi (Ed.), *Design Innovations for Contemporary Interiors and Civic Art* (pp. 348–367). Hershey, PA: IGI Global. doi:10.4018/978-1-5225-0666-9.ch019

Bianchini, C., Ippolito, A., & Bartolomei, C. (2015). The Surveying and Representation Process Applied to Architecture: Non-Contact Methods for the Documentation of Cultural Heritage. In S. Brusaporci (Ed.), *Handbook of Research on Emerging Digital Tools for Architectural Surveying, Modeling, and Representation* (pp. 44–93). Hershey, PA: IGI Global. doi:10.4018/978-1-4666-8379-2.ch002

Bianconi, F., Filippucci, M., Margutti, M., & Stramaccia, M. (2018). Drawing Tensegrity, Discover Trans-Polyhedra: From Polyhedra and Tessellation to Find New Architectural Forms. In D. D'Uva (Ed.), *Handbook of Research on Form and Morphogenesis in Modern Architectural Contexts* (pp. 41–68). Hershey, PA: IGI Global. doi:10.4018/978-1-5225-3993-3.ch003

Bobo, T., & Nechena, H. (2018). Fighting for Existence and Recognition Among Sub-Dynasty Communities: A Case Study of the Nerumedzo People in Zimbabwe. In I. Rosa, J. Lopes, R. Ribeiro, & A. Mendes (Eds.), *Handbook of Research on Methods and Tools for Assessing Cultural Landscape Adaptation* (pp. 53–71). Hershey, PA: IGI Global. doi:10.4018/978-1-5225-4186-8.ch003

Brusaporci, S. (2015). On Visual Computing for Architectural Heritage. In S. Brusaporci (Ed.), *Handbook of Research on Emerging Digital Tools for Architectural Surveying, Modeling, and Representation* (pp. 94–123). Hershey, PA: IGI Global. doi:10.4018/978-1-4666-8379-2.ch003

Buffi, A., Filippucci, M., & Bianconi, F. (2018). Optimization and Evolution in Architectural Morphogenesis: Evolutionary Principles Applied to Mass Housing. In D. D'Uva (Ed.), *Handbook of Research on Form and Morphogenesis in Modern Architectural Contexts* (pp. 263–282). Hershey, PA: IGI Global. doi:10.4018/978-1-5225-3993-3.ch013

Caeiro, M. (2017). A Critical Rhetoric for Urban Innovation: Art Staging the Civic. In L. Crespi (Ed.), *Design Innovations for Contemporary Interiors and Civic Art* (pp. 102–124). Hershey, PA: IGI Global. doi:10.4018/978-1-5225-0666-9.ch005

Cañas-Bajo, J., & Silvennoinen, J. (2018). Experiencing Commercial Videos for Online Shopping: A Cross-Cultural User's Design Approach. In M. Khosrow-Pour, D.B.A. (Ed.), Enhancing Art, Culture, and Design With Technological Integration(pp. 183-214). Hershey, PA: IGI Global. doi:10.4018/978-1-5225-5023-5.ch010

Casu, P., & Pisu, C. (2015). 3D Reconstruction for the Interpretation of Partly Lost or Never Accomplished Architectural Heritage. In S. Brusaporci (Ed.), *Handbook of Research on Emerging Digital Tools for Architectural Surveying, Modeling, and Representation* (pp. 686–721). Hershey, PA: IGI Global. doi:10.4018/978-1-4666-8379-2.ch023

Cattaneo, T., De Lotto, R., & Venco, E. M. (2016). Functional Reuse and Intensification of Rural-Urban Context: Rural Architectural Urbanism. [IJAEIS]. *International Journal of Agricultural and Environmental Information Systems*, *7*(1), 1–27. doi:10.4018/IJAEIS.2016010101

Chiarella, M., & Costa, M. D. (2018). Dynamic Generative Systems: Simple Parametric Strategies for Complex Architectural Objects and Spaces. In D. D'Uva (Ed.), *Handbook of Research on Form and Morphogenesis in Modern Architectural Contexts* (pp. 108–128). Hershey, PA: IGI Global. doi:10.4018/978-1-5225-3993-3.ch006

Chiavoni, E. (2015). Integrated Methodologies for the Study and Documentation of the Architectural and Archaeological Heritage: The Treasury of Petra in Jordan. In S. Brusaporci (Ed.), *Handbook of Research on Emerging Digital Tools for Architectural Surveying, Modeling, and Representation* (pp. 751–787). Hershey, PA: IGI Global. doi:10.4018/978-1-4666-8379-2.ch025

Cigola, M. (2015). Digital Tools for Urban and Architectural Heritage. In S. Brusaporci (Ed.), *Handbook of Research on Emerging Digital Tools for Architectural Surveying, Modeling, and Representation* (pp. 403–424). Hershey, PA: IGI Global. doi:10.4018/978-1-4666-8379-2.ch014

Clini, P., Quattrini, R., Frontoni, E., & Nespeca, R. (2015). A New Cloud Library for Integrated Surveys: The Ancient Via Flaminia and the Nextone Project. In S. Brusaporci (Ed.), *Handbook of Research on Emerging Digital Tools for Architectural Surveying, Modeling, and Representation* (pp. 579–606). Hershey, PA: IGI Global. doi:10.4018/978-1-4666-8379-2.ch020

Colonnese, F. (2017). Human Figure as a Cultural Mediator in Architectural Drawings. In G. Koç, M. Claes, & B. Christiansen (Eds.), *Cultural Influences on Architecture* (pp. 90–129). Hershey, PA: IGI Global. doi:10.4018/978-1-5225-1744-3.ch004

Continenza, R., & Trizio, I. (2015). The SIArch-Univaq: An Architectural Information System for Cultural Heritage. In S. Brusaporci (Ed.), *Handbook of Research on Emerging Digital Tools for Architectural Surveying, Modeling, and Representation* (pp. 311–336). Hershey, PA: IGI Global. doi:10.4018/978-1-4666-8379-2.ch010

Converso, S. (2018). The Fabrication of "The Cloud": Introducing Material Constraints in Complex Geometry Optimization. In D. D'Uva (Ed.), *Handbook of Research on Form and Morphogenesis in Modern Architectural Contexts* (pp. 381–403). Hershey, PA: IGI Global. doi:10.4018/978-1-5225-3993-3.ch018

Crespi, G. (2017). Art and Space: New Boundaries of Intervention. In L. Crespi (Ed.), *Design Innovations for Contemporary Interiors and Civic Art* (pp. 125–137). Hershey, PA: IGI Global. doi:10.4018/978-1-5225-0666-9.ch006

Crespi, L. (2017). Design Strategies as Paradigms. In L. Crespi (Ed.), *Design Innovations for Contemporary Interiors and Civic Art* (pp. 26–64). Hershey, PA: IGI Global. doi:10.4018/978-1-5225-0666-9.ch002

Crespi, L. (2017). Designing Interiors: A Guide for Contemporary Interiors Landscape Design. In L. Crespi (Ed.), *Design Innovations for Contemporary Interiors and Civic Art* (pp. 1–25). Hershey, PA: IGI Global. doi:10.4018/978-1-5225-0666-9.ch001

D'Uva, D. (2018). Geometrical Characterization of a Rototraslative Generative Tower: The Turning Torso Case Study. In D. D'Uva (Ed.), *Handbook of Research on Form and Morphogenesis in Modern Architectural Contexts* (pp. 219–242). Hershey, PA: IGI Global. doi:10.4018/978-1-5225-3993-3.ch011

Das, K. K., & Sharma, N. K. (2016). Post Disaster Housing Management for Sustainable Urban Development: A Review. [IJGEE]. *International Journal of Geotechnical Earthquake Engineering*, *7*(1), 1–18. doi:10.4018/IJGEE.2016010101

Di Prete, B. (2017). Urban Interior Design: Strategies for Public Living Spaces. In L. Crespi (Ed.), *Design Innovations for Contemporary Interiors and Civic Art* (pp. 156–185). Hershey, PA: IGI Global. doi:10.4018/978-1-5225-0666-9.ch008

Di Sabatino, P. A. (2017). SoftAssertions: A Progressive Paradigm for Urban Cultural Heritage, Interior Urbanism, and Interior Design. In L. Crespi (Ed.), *Design Innovations for Contemporary Interiors and Civic Art* (pp. 65–86). Hershey, PA: IGI Global. doi:10.4018/978-1-5225-0666-9.ch003

Dore, C., & Murphy, M. (2015). Historic Building Information Modelling (HBIM). In S. Brusaporci (Ed.), *Handbook of Research on Emerging Digital Tools for Architectural Surveying, Modeling, and Representation* (pp. 233–273). Hershey, PA: IGI Global. doi:10.4018/978-1-4666-8379-2.ch007

Doyle, D. (2018). Imagination and the Phenomenology of Virtual Practice. In M. Khosrow-Pour, D.B.A. (Ed.), Enhancing Art, Culture, and Design With Technological Integration (pp. 131-151). Hershey, PA: IGI Global. doi:10.4018/978-1-5225-5023-5.ch007

Eilouti, B. H. (2018). Concept as the DNA for Morphogenesis: A Case Study of Contemporary Architecture. In D. D'Uva (Ed.), *Handbook of Research on Form and Morphogenesis in Modern Architectural Contexts* (pp. 283–309). Hershey, PA: IGI Global. doi:10.4018/978-1-5225-3993-3.ch014

Ekman, U. (2018). Smart City Planning: Complexity. [IJEPR]. *International Journal of E-Planning Research*, *7*(3), 1–21. doi:10.4018/IJEPR.2018070101

Empler, T. (2018). Representation in Architecture as Idea, Physical Model, 3D Modeling, BIM. In D. D'Uva (Ed.), *Handbook of Research on Form and Morphogenesis in Modern Architectural Contexts* (pp. 93–107). Hershey, PA: IGI Global. doi:10.4018/978-1-5225-3993-3.ch005

Erioli, A. (2018). Anexact Paths: Computation, Continuity, and Tectonics in the Design Process. In D. D'Uva (Ed.), *Handbook of Research on Form and Morphogenesis in Modern Architectural Contexts* (pp. 1–19). Hershey, PA: IGI Global. doi:10.4018/978-1-5225-3993-3.ch001

Fallacara, G., & Barberio, M. (2018). Parametric Morphogenesis, Robotic Fabrication, and Construction of Novel Stereotomic Hypar Morphologies: Hypar Gate, Hypar Wall, and Hypar Vault. In D. D'Uva (Ed.), *Handbook of Research on Form and Morphogenesis in Modern Architectural Contexts* (pp. 329–353). Hershey, PA: IGI Global. doi:10.4018/978-1-5225-3993-3.ch016

Fatta, F. (2015). Communication, Technology, and Digital Culture for the Conservation and Enhancement of the Architectural Heritage. In S. Brusaporci (Ed.), *Handbook of Research on Emerging Digital Tools for Architectural Surveying, Modeling, and Representation* (pp. 446–475). Hershey, PA: IGI Global. doi:10.4018/978-1-4666-8379-2.ch016

Ferrighi, A. (2015). Cities over Space and Time: Historical GIS for Urban History. In S. Brusaporci (Ed.), *Handbook of Research on Emerging Digital Tools for Architectural Surveying, Modeling, and Representation* (pp. 425–445). Hershey, PA: IGI Global. doi:10.4018/978-1-4666-8379-2.ch015

Franchina, A., Maggio, F., & Vattano, S. (2015). Female Architecture: Unbuilt Digital Archive. In S. Brusaporci (Ed.), *Handbook of Research on Emerging Digital Tools for Architectural Surveying, Modeling, and Representation* (pp. 510–549). Hershey, PA: IGI Global. doi:10.4018/978-1-4666-8379-2.ch018

Gaiani, M. (2015). Color Acquisition, Management, Rendering, and Assessment in 3D Reality-Based Models Construction. In S. Brusaporci (Ed.), *Handbook of Research on Emerging Digital Tools for Architectural Surveying, Modeling, and Representation* (pp. 1–43). Hershey, PA: IGI Global. doi:10.4018/978-1-4666-8379-2.ch001

Galizia, M., Lo Faro, A., & Santagati, C. (2015). Understanding and Conserving Fortified and Difficult to Access Architectures through Digital Survey: A Case History in Southern Sicily. In S. Brusaporci (Ed.), *Handbook of Research on Emerging Digital Tools for Architectural Surveying, Modeling, and Representation* (pp. 654–685). Hershey, PA: IGI Global. doi:10.4018/978-1-4666-8379-2.ch022

Garagnani, S. (2015). Semantic Representation of Accurate Surveys for the Cultural Heritage: BIM Applied to the Existing Domain. In S. Brusaporci (Ed.), *Handbook of Research on Emerging Digital Tools for Architectural Surveying, Modeling, and Representation* (pp. 292–310). Hershey, PA: IGI Global. doi:10.4018/978-1-4666-8379-2.ch009

Garfella, J. T., Máñez, M. J., & Martínez, J. Á. (2015). Comparative Study of Graphic Representation Methods on Architectural Heritage. In S. Brusaporci (Ed.), *Handbook of Research on Emerging Digital Tools for Architectural Surveying, Modeling, and Representation* (pp. 182–209). Hershey, PA: IGI Global. doi:10.4018/978-1-4666-8379-2.ch005

Giostra, S. (2018). Energy/Form/Rule: Experiments in Energy Form-Finding. In D. D'Uva (Ed.), *Handbook of Research on Form and Morphogenesis in Modern Architectural Contexts* (pp. 404–431). Hershey, PA: IGI Global. doi:10.4018/978-1-5225-3993-3.ch019

Gonçalves, C. (2018). Models, Methods, and Metrics to Measure Socioeconomic Resilience: Two Portuguese Urban Systems as Case Studies. In I. Rosa, J. Lopes, R. Ribeiro, & A. Mendes (Eds.), *Handbook of Research on Methods and Tools for Assessing Cultural Landscape Adaptation* (pp. 346–367). Hershey, PA: IGI Global. doi:10.4018/978-1-5225-4186-8.ch014

Gou, Z. (2017). Workplace Design Revolution: The Inside-Out Urbanism. In L. Crespi (Ed.), *Design Innovations for Contemporary Interiors and Civic Art* (pp. 225–240). Hershey, PA: IGI Global. doi:10.4018/978-1-5225-0666-9.ch012

Gourdoukis, D. (2018). The Future is Analog: A Post-Protocological Approach to the Production of Form in Architecture. In D. D'Uva (Ed.), *Handbook of Research on Form and Morphogenesis in Modern Architectural Contexts* (pp. 191–218). Hershey, PA: IGI Global. doi:10.4018/978-1-5225-3993-3.ch010

Hafiz, D. (2017). Improving Interior-Design Decision-Making in Daylit Spaces: A Case Study. In L. Crespi (Ed.), *Design Innovations for Contemporary Interiors and Civic Art* (pp. 368–388). Hershey, PA: IGI Global. doi:10.4018/978-1-5225-0666-9.ch020

Horta, M. B., Cabral, M. I., Pires, I., Bachi, L. S., Luz, A., Fernandes, G. W., ... Carvalho-Ribeiro, S. (2018). Assessing Urban Ecosystem Services: Different Methodological Approaches Applied in Brazil, Germany, and Portugal. In I. Rosa, J. Lopes, R. Ribeiro, & A. Mendes (Eds.), *Handbook of Research on Methods and Tools for Assessing Cultural Landscape Adaptation* (pp. 183–220). Hershey, PA: IGI Global. doi:10.4018/978-1-5225-4186-8.ch008

Ippoliti, E., Meschini, A., & Sicuranza, F. (2015). Digital Photogrammetry and Structure from Motion for Architectural Heritage: Comparison and Integration between Procedures. In S. Brusaporci (Ed.), *Handbook of Research on Emerging Digital Tools for Architectural Surveying, Modeling, and Representation* (pp. 124–181). Hershey, PA: IGI Global. doi:10.4018/978-1-4666-8379-2.ch004

Johansson, M. (2018). Soundscaping. In M. Khosrow-Pour, D.B.A. (Ed.), Enhancing Art, Culture, and Design With Technological Integration (pp. 169-182). Hershey, PA: IGI Global. doi:10.4018/978-1-5225-5023-5.ch009

Karnaukhova, O. (2017). “Bridges” and “Gaps” on Maps of Multicultural Cities: The Story of the South Russian Agglomeration. In G. Koç, M. Claes, & B. Christiansen (Eds.), *Cultural Influences on Architecture* (pp. 181–199). Hershey, PA: IGI Global. doi:10.4018/978-1-5225-1744-3.ch007

Kasemsap, K. (2017). The Fundamentals of Social Capital. In G. Koç, M. Claes, & B. Christiansen (Eds.), *Cultural Influences on Architecture* (pp. 259–292). Hershey, PA: IGI Global. doi:10.4018/978-1-5225-1744-3.ch010

Koç, G., & Christiansen, B. (2017). The Role of Cultural Indoctrination in Architectural Style: Religion as a Mediator. In G. Koç, M. Claes, & B. Christiansen (Eds.), *Cultural Influences on Architecture* (pp. 1–34). Hershey, PA: IGI Global. doi:10.4018/978-1-5225-1744-3.ch001

Köroğlu, C. Z., & Köroğlu, M. A. (2018). On the Transformation of Social Movements: An Analysis From the East-West Axis. In M. Khosrow-Pour, D.B.A. (Ed.), Enhancing Art, Culture, and Design With Technological Integration (pp. 50-74). Hershey, PA: IGI Global. doi:10.4018/978-1-5225-5023-5.ch003

Kupritz, V. W. (2017). The Communicative Nature of Space in Organizations. In G. Koç, M. Claes, & B. Christiansen (Eds.), *Cultural Influences on Architecture* (pp. 58–89). Hershey, PA: IGI Global. doi:10.4018/978-1-5225-1744-3.ch003

La Mantia, M. (2015). A Virtual Journey through 2D and 3D Elaborations Recorded with Range-Based and Image-Based Method: The Experience of Vitelleschi Palace in Tarquinia. In S. Brusaporci (Ed.), *Handbook of Research on Emerging Digital Tools for Architectural Surveying, Modeling, and Representation* (pp. 607–653). Hershey, PA: IGI Global. doi:10.4018/978-1-4666-8379-2.ch021

Lavrador, A., & Rocha, J. (2018). The Role of Landscape in the Representation of Portuguese Wine Producing Regions. In I. Rosa, J. Lopes, R. Ribeiro, & A. Mendes (Eds.), *Handbook of Research on Methods and Tools for Assessing Cultural Landscape Adaptation* (pp. 276–298). Hershey, PA: IGI Global. doi:10.4018/978-1-5225-4186-8.ch011

Lo Turco, M., & Pagliero, Y. B. (2018). Representation, Design, and Management of Generative Algorithms for Architecture: The Case Study of the Institute of the Arab World, in Paris. In D. D'Uva (Ed.), Handbook of Research on Form and Morphogenesis in Modern Architectural Contexts (pp. 243-262). Hershey, PA: IGI Global. doi:10.4018/978-1-5225-3993-3.ch012

Lourenço, A. C. (2018). City Landscape: Confluence Between Ecological Conditions and Urban Morphology in the City of Lisbon. In I. Rosa, J. Lopes, R. Ribeiro, & A. Mendes (Eds.), *Handbook of Research on Methods and Tools for Assessing Cultural Landscape Adaptation* (pp. 368–395). Hershey, PA: IGI Global. doi:10.4018/978-1-5225-4186-8.ch015

Louro, M., & Oliveira, F. (2018). In Search of a Lost Authenticity: Tourism, Renaturation, and City – The Case of the Tagus Estuary. In I. Rosa, J. Lopes, R. Ribeiro, & A. Mendes (Eds.), *Handbook of Research on Methods and Tools for Assessing Cultural Landscape Adaptation* (pp. 72–89). Hershey, PA: IGI Global. doi:10.4018/978-1-5225-4186-8.ch004

Luigini, A., & Vattano, S. (2018). Ten Words About Morphogenetic Images: A Discreet Pathway Between Science, Art, and Architecture. In D. D'Uva (Ed.), *Handbook of Research on Form and Morphogenesis in Modern Architectural Contexts* (pp. 129–151). Hershey, PA: IGI Global. doi:10.4018/978-1-5225-3993-3.ch007

Mayo, S. (2018). A Collective Consciousness Model in a Post-Media Society. In M. Khosrow-Pour, D.B.A. (Ed.), Enhancing Art, Culture, and Design With Technological Integration (pp. 25-49). Hershey, PA: IGI Global. doi:10.4018/978-1-5225-5023-5.ch002

McIntosh, J., Marques, B., & Hatton, W. (2018). Indigenous Cultural Knowledge for Therapeutic Landscape Design. In I. Rosa, J. Lopes, R. Ribeiro, & A. Mendes (Eds.), *Handbook of Research on Methods and Tools for Assessing Cultural Landscape Adaptation* (pp. 28–52). Hershey, PA: IGI Global. doi:10.4018/978-1-5225-4186-8.ch002

Megchun, B. I. (2017). Retail Design Brand Experiences through Merging Art, Design, and Science. In L. Crespi (Ed.), *Design Innovations for Contemporary Interiors and Civic Art* (pp. 209–224). Hershey, PA: IGI Global. doi:10.4018/978-1-5225-0666-9.ch011

Meloni, P. (2017). New Domestic Rituals: Note on Design Anthropology. In L. Crespi (Ed.), *Design Innovations for Contemporary Interiors and Civic Art* (pp. 87–101). Hershey, PA: IGI Global. doi:10.4018/978-1-5225-0666-9.ch004

Murialdo, F. (2017). Practice of Consumption and Spaces for Goods. In L. Crespi (Ed.), *Design Innovations for Contemporary Interiors and Civic Art* (pp. 190–208). Hershey, PA: IGI Global. doi:10.4018/978-1-5225-0666-9.ch010

Naboni, R., & Paoletti, I. (2018). Architectural Morphogenesis Through Topology Optimization. In D. D'Uva (Ed.), *Handbook of Research on Form and Morphogenesis in Modern Architectural Contexts* (pp. 69–92). Hershey, PA: IGI Global. doi:10.4018/978-1-5225-3993-3.ch004

Nicolin, P. (2017). Urban Farming Movement. In L. Crespi (Ed.), *Design Innovations for Contemporary Interiors and Civic Art* (pp. 186–189). Hershey, PA: IGI Global. doi:10.4018/978-1-5225-0666-9.ch009

Olalere, F. E. (2018). Experience-Centred Design and the Role of Computer-Aided Tools in the Creative Process. In M. Khosrow-Pour, D.B.A. (Ed.), Enhancing Art, Culture, and Design With Technological Integration (pp. 99-113). Hershey, PA: IGI Global. doi:10.4018/978-1-5225-5023-5.ch005

Olalere, F. E. (2018). Rapid Product Development and Application in Ceramics Production. In M. Khosrow-Pour, D.B.A. (Ed.), Enhancing Art, Culture, and Design With Technological Integration (pp. 215-233). Hershey, PA: IGI Global. doi:10.4018/978-1-5225-5023-5.ch011

Oliveira, M. D., & Ribeiro, J. T. (2018). Industrial Heritage as an Operative Territorial Resource: Cultural Landscape of Alentejo Pyrite. In I. Rosa, J. Lopes, R. Ribeiro, & A. Mendes (Eds.), *Handbook of Research on Methods and Tools for Assessing Cultural Landscape Adaptation* (pp. 153–181). Hershey, PA: IGI Global. doi:10.4018/978-1-5225-4186-8.ch007

Onay, N. S., Garip, B., & Teixeira, M. B. (2017). Searching for Flexible Interior Design Solutions: Mass Housing Units in Istanbul. In L. Crespi (Ed.), *Design Innovations for Contemporary Interiors and Civic Art* (pp. 270–284). Hershey, PA: IGI Global. doi:10.4018/978-1-5225-0666-9.ch015

Osello, A., Acquaviva, A., Dalmasso, D., Erba, D., Del Giudice, M., Macii, E., & Patti, E. (2015). BIM and Interoperability for Cultural Heritage through ICT. In S. Brusaporci (Ed.), *Handbook of Research on Emerging Digital Tools for Architectural Surveying, Modeling, and Representation* (pp. 274–291). Hershey, PA: IGI Global. doi:10.4018/978-1-4666-8379-2.ch008

Paris, L. (2015). Shape and Geometry in the Integrated Digital Survey. In S. Brusaporci (Ed.), *Handbook of Research on Emerging Digital Tools for Architectural Surveying, Modeling, and Representation* (pp. 210–232). Hershey, PA: IGI Global. doi:10.4018/978-1-4666-8379-2.ch006

Pelliccio, A., Ottaviano, E., & Rea, P. (2015). Digital and Mechatronic Technologies Applied to the Survey of Brownfields. In S. Brusaporci (Ed.), *Handbook of Research on Emerging Digital Tools for Architectural Surveying, Modeling, and Representation* (pp. 813–829). Hershey, PA: IGI Global. doi:10.4018/978-1-4666-8379-2.ch027

Pineda, R. G. (2018). Remediating Interaction: Towards a Philosophy of Human-Computer Relationship. In M. Khosrow-Pour, D.B.A. (Ed.), Enhancing Art, Culture, and Design With Technological Integration (pp. 75-98). Hershey, PA: IGI Global. doi:10.4018/978-1-5225-5023-5.ch004

Rahm, P. (2017). The Anthropocene Style: Towards a New Decorative Style. In L. Crespi (Ed.), *Design Innovations for Contemporary Interiors and Civic Art* (pp. 258–269). Hershey, PA: IGI Global. doi:10.4018/978-1-5225-0666-9.ch014

Randa, I. O. (2016). Integrated Model of Affordable Housing Delivery for the City of Windhoek. [IJCESC]. *International Journal of Civic Engagement and Social Change*, *3*(3), 1–15. doi:10.4018/IJCESC.2016070101

Ribeiro, R. J., Lopes, J. C., & Boucault, F. (2018). Alternative Tool for an Integrative Landscape Interpretation: Case Study of the Arrábida Maritime Coast, Portugal. In I. Rosa, J. Lopes, R. Ribeiro, & A. Mendes (Eds.), *Handbook of Research on Methods and Tools for Assessing Cultural Landscape Adaptation* (pp. 299–322). Hershey, PA: IGI Global. doi:10.4018/978-1-5225-4186-8.ch012

Robertson, S. (2017). Between the Planned and the Lived City: New Road, Brighton Drawn. In L. Crespi (Ed.), *Design Innovations for Contemporary Interiors and Civic Art* (pp. 139–155). Hershey, PA: IGI Global. doi:10.4018/978-1-5225-0666-9.ch007

Romero, C. S. (2018). Culture Industry: A Contemporary Revision of the Term of Theodor Adorno. In M. Khosrow-Pour, D.B.A. (Ed.), Enhancing Art, Culture, and Design With Technological Integration (pp. 1-24). Hershey, PA: IGI Global. doi:10.4018/978-1-5225-5023-5.ch001

Rosa, I. S., Lopes, J. C., & Ribeiro, R. J. (2015). The Landscape Cultural Construction: A Recognition of the Roman Tradition. In S. Brusaporci (Ed.), *Handbook of Research on Emerging Digital Tools for Architectural Surveying, Modeling, and Representation* (pp. 788–812). Hershey, PA: IGI Global. doi:10.4018/978-1-4666-8379-2.ch026

Russo, M., & Manferdini, A. M. (2015). Integrated Multi-Scalar Approach for 3D Cultural Heritage Acquisitions. In S. Brusaporci (Ed.), *Handbook of Research on Emerging Digital Tools for Architectural Surveying, Modeling, and Representation* (pp. 337–360). Hershey, PA: IGI Global. doi:10.4018/978-1-4666-8379-2.ch011

Ruttico, P. (2018). Innovative Construction Systems: New Design Aesthetics Through Competitive Realization Processes. In D. D'Uva (Ed.), *Handbook of Research on Form and Morphogenesis in Modern Architectural Contexts* (pp. 354–379). Hershey, PA: IGI Global. doi:10.4018/978-1-5225-3993-3.ch017

Shakibamanesh, A., & Ghorbanian, M. (2015). Using Advanced Approaches in Urban Design Researches: A Mutation from 3D Digital Models to Virtual Reality. In S. Brusaporci (Ed.), *Handbook of Research on Emerging Digital Tools for Architectural Surveying, Modeling, and Representation* (pp. 722–750). Hershey, PA: IGI Global. doi:10.4018/978-1-4666-8379-2.ch024

Silva, C. F. (2018). Time Operations. In I. Rosa, J. Lopes, R. Ribeiro, & A. Mendes (Eds.), *Handbook of Research on Methods and Tools for Assessing Cultural Landscape Adaptation* (pp. 90–121). Hershey, PA: IGI Global. doi:10.4018/978-1-5225-4186-8.ch005

Silva, C. F., & Fernandes, M. C. (2018). The Cycles of Impermanent Alterity in Nazaré. In I. Rosa, J. Lopes, R. Ribeiro, & A. Mendes (Eds.), *Handbook of Research on Methods and Tools for Assessing Cultural Landscape Adaptation* (pp. 434–465). Hershey, PA: IGI Global. doi:10.4018/978-1-5225-4186-8.ch018

Sirago, V. (2018). Stadium Design in Digital Age: Automatic Bowl Generation Tools. In D. D'Uva (Ed.), *Handbook of Research on Form and Morphogenesis in Modern Architectural Contexts* (pp. 432–463). Hershey, PA: IGI Global. doi:10.4018/978-1-5225-3993-3.ch020

Spallone, R. (2015). Digital Reconstruction of Demolished Architectural Masterpieces, 3D Modeling, and Animation: The Case Study of Turin Horse Racing by Mollino. In S. Brusaporci (Ed.), *Handbook of Research on Emerging Digital Tools for Architectural Surveying, Modeling, and Representation* (pp. 476–509). Hershey, PA: IGI Global. doi:10.4018/978-1-4666-8379-2.ch017

Symeonidou, I. (2018). Digital Creativity: Embracing New Technologies for Architectural Innovation. In D. D'Uva (Ed.), *Handbook of Research on Form and Morphogenesis in Modern Architectural Contexts* (pp. 175–189). Hershey, PA: IGI Global. doi:10.4018/978-1-5225-3993-3.ch009

Torsi, S. (2017). City of Beats: Analysing Flânerie as a Practice for Living the Physical Space. In G. Koç, M. Claes, & B. Christiansen (Eds.), *Cultural Influences on Architecture* (pp. 157–180). Hershey, PA: IGI Global. doi:10.4018/978-1-5225-1744-3.ch006

Tran, B. (2017). Psychological (and Emotional) Architecture: The Values and Benefits of Nature-Based Architecture – Biophilia. In G. Koç, M. Claes, & B. Christiansen (Eds.), *Cultural Influences on Architecture* (pp. 200–230). Hershey, PA: IGI Global. doi:10.4018/978-1-5225-1744-3.ch008

Tran, B. (2017). The Architect of Organizational Psychology: The Geert Hofstede's Dimensions of Cultural (Corporate and Organizational) Identity. In G. Koç, M. Claes, & B. Christiansen (Eds.), *Cultural Influences on Architecture* (pp. 231–258). Hershey, PA: IGI Global. doi:10.4018/978-1-5225-1744-3.ch009

Trentin, L. (2017). The SPACE Between: On the Relationship between Drawing and Design. In L. Crespi (Ed.), *Design Innovations for Contemporary Interiors and Civic Art* (pp. 307–318). Hershey, PA: IGI Global. doi:10.4018/978-1-5225-0666-9.ch017

Uzun, F. V., & Somuncu, M. (2018). Cultural Landscape: An Evaluation From Past to Present. In I. Rosa, J. Lopes, R. Ribeiro, & A. Mendes (Eds.), *Handbook of Research on Methods and Tools for Assessing Cultural Landscape Adaptation* (pp. 1–27). Hershey, PA: IGI Global. doi:10.4018/978-1-5225-4186-8.ch001

Vanossi, A. (2018). Interaction and Intersection Between Digital Modelling and Design in Architecture: Different Approaches in Parametric Design. In D. D'Uva (Ed.), *Handbook of Research on Form and Morphogenesis in Modern Architectural Contexts* (pp. 152–174). Hershey, PA: IGI Global. doi:10.4018/978-1-5225-3993-3.ch008

Vatan, M. (2017). Evolution of Construction Systems: Cultural Effects On Traditional Structures and Their Reflection On Modern Building Construction. In G. Koç, M. Claes, & B. Christiansen (Eds.), *Cultural Influences on Architecture* (pp. 35–57). Hershey, PA: IGI Global. doi:10.4018/978-1-5225-1744-3.ch002

Vaz de Freitas, I., Marques, J., Rodrigues, C. A., & Sousa, C. (2018). Urban Landscape Quality Management and Monitoring: A Methodological Proposal to Study the Case of Porto, Portugal. In I. Rosa, J. Lopes, R. Ribeiro, & A. Mendes (Eds.), *Handbook of Research on Methods and Tools for Assessing Cultural Landscape Adaptation* (pp. 396–413). Hershey, PA: IGI Global. doi:10.4018/978-1-5225-4186-8.ch016

Voyatzaki, M., & Gourdoukis, D. (2018). From Morphology to Morphogenesis: On Speculative Architectural Design Pedagogy. In D. D'Uva (Ed.), *Handbook of Research on Form and Morphogenesis in Modern Architectural Contexts* (pp. 20–40). Hershey, PA: IGI Global. doi:10.4018/978-1-5225-3993-3.ch002

Yücel, G. F., Işık, B., & Cömert, N. Z. (2018). An Ecological Assessment Analysis: The Kanlidere River in North Cyprus. In I. Rosa, J. Lopes, R. Ribeiro, & A. Mendes (Eds.), *Handbook of Research on Methods and Tools for Assessing Cultural Landscape Adaptation* (pp. 414–433). Hershey, PA: IGI Global. doi:10.4018/978-1-5225-4186-8.ch017

Zhu, X., Galland, P. L., Dick, R. D., & Wallis, R. K. (2017). Towards Regenerative Urban Environments: Urban Interiors for the Sustainable Human Species. In L. Crespi (Ed.), *Design Innovations for Contemporary Interiors and Civic Art* (pp. 319–347). Hershey, PA: IGI Global. doi:10.4018/978-1-5225-0666-9.ch018

About the Authors

Gebrail Bekdaş, Associative Professor, is researcher in Structural Control and Optimization at Istanbul University - Cerrahpaşa. He obtained his DPhil in Structural Engineering from Istanbul University with a thesis subject of design of cylindrical walls. He co-organized 15th EU-ME Workshop: Metaheuristic and Engineering in Istanbul. In optimization, he organized several mini-symposiums or special sections in prestigious international events such as 12th International Congress on Mechanics (HSTAM2019), 9th International Congress on Computational Mechanics (GRACM 2018), the Biennial International Conference on Engineering Vibration (ICoEV-2015), 3rd International Conference on Multiple Criteria Decision Making (MCDM 2015), 11th. World Congress on Computational Mechanics (WCCM2014), International Conference on Engineering and Applied Sciences Optimization (OPTI 2014), 11th Biennial International Conference on Vibration Problems (ICOVP2013), 3rd European Conference of Civil Engineering and 10th, 11th and 15th International Conference of Numerical Analysis and Applied Mathematics (ICNAAM). He co-edited Metaheuristics and Optimization in Civil Engineering published by Springer and he is one of the guest editors in 2017 special issue of KSCE Journal of Civil Engineering. He has authored about 250 papers for journals and scientific events.

Aylin Ece Kayabekir is a student of civil engineering PhD program of Istanbul University - Cerrahpaşa. She is working on structural optimization and control. Up to now, the authored more than 30 papers about optimization of RC retaining walls, RC shear walls, frames, carbon fiber reinforced polymers, dynamic of structures interacting with soil and structural control in selected journals covered in SCI and international conferences. In optimization, she also organized several mini-symposiums or special sections in prestigious international events 12th International Congress on Mechanics (HSTAM2019) and 9th International Congress on Computational Mechanics

(GRACM 2018). Now, she is working on a book project about optimization of reinforced concrete structures. She is supported by Turkish High Education Council and The Scientific and Technological Research Council of Turkey because of the importance of her thesis subject.

Sinan Melih Nigdeli, Associative Professor, is researcher in Structural Control and Optimization at Istanbul University - Cerrahpaşa. He obtained his DPhil in Structural Engineering from Istanbul Technical University with a thesis subject of active control. He co-organized 15th EU-ME Workshop: Metaheuristic and Engineering in Istanbul. In optimization, he organized several mini-symposiums or special sections in prestigious international events such as 12th International Congress on Mechanics (HSTAM2019), 9th International Congress on Computational Mechanics (GRACM 2018), the Biennial International Conference on Engineering Vibration (ICoEV-2015), 3rd International Conference on Multiple Criteria Decision Making (MCDM 2015), 11th. World Congress on Computational Mechanics (WCCM2014), International Conference on Engineering and Applied Sciences Optimization (OPTI 2014), 11th Biennial International Conference on Vibration Problems (ICOVP2013), 3rd European Conference of Civil Engineering and 10th, 11th and 15th International Conference of Numerical Analysis and Applied Mathematics (ICNAAM). He co-edited Metaheuristics and Optimization in Civil Engineering published by Springer and he is one of the guest editors in 2017 special issue of KSCE Journal of Civil Engineering. He has authored about 250 papers for journals and scientific events.

Index

www.ingramcontent.com/pod-product-compliance
Lightning Source LLC
LaVergne TN
LVHW061535290926
849162LV00004B/797